Zeolites - New Challenges

Edited by Karmen Margeta
and Anamarija Farkaš

Published in London, United Kingdom

IntechOpen

Supporting open minds since 2005

Zeolites - New Challenges
http://dx.doi.org/10.5772/intechopen.77482
Edited by Karmen Margeta and Anamarija Farkaš

Contributors
Nandini Das, Jugal Kishore Das, Peng Wu, Hao Xu, Adalgisa Tavolaro, Silvia Catalano, Palmira Tavolaro, Jacques Fraissard, Mykhaylo Petryk, Mykola Ivanchov, Daniel Canet, Sebastien Leclerc, Yashonath Subramanian, Shubhadeep Nag, Camelia Tulcan, Simona Marc, Karmen Margeta, Anamarija Farkaš

First published in London, United Kingdom, 2020 by IntechOpen
IntechOpen is the global imprint of INTECHOPEN LIMITED, registered in England and Wales, registration number: 11086078, 7th floor, 10 Lower Thames Street, London, EC3R 6AF, United Kingdom
Printed in Croatia

British Library Cataloguing-in-Publication Data
A catalogue record for this book is available from the British Library

Additional hard and PDF copies can be obtained from orders@intechopen.com

Zeolites - New Challenges
Edited by Karmen Margeta and Anamarija Farkaš
p. cm.
Print ISBN 978-1-78985-469-5
Online ISBN 978-1-78985-470-1
eBook (PDF) ISBN 978-1-83880-850-1

For EU product safety concerns:
IN TECH d.o.o., Prolaz Marije Krucifikse Kozulić 3, 51000 Rijeka, Croatia, info@intechopen.com or visit our website at intechopen.com.

Meet the editors

Dr. Karmen Margeta (editor) is a Senior Scientist with a BSc, MSc, and PhD in Chemical Engineering from the University of Zagreb, Croatia, where she works in the field of water treatment and zeolite materials. She is an expert in the fields of chemical (analytical) testing and environmental testing. She is involved in applied research related to material science for energy technologies and water treatment technologies. She has authored and co-authored more than 70 journal and conference papers and several book chapters. In 2017 she published *Seawater Steam Engine as a "Prime Mover" for the Third Industrial Revolution*, a book that describes a radical new technology for stopping climate change. In 2015 Dr. Margeta was a member of the American Association of Science and Technology. She is also inventor of international patent WO/2013/072709. 2017 she published a book titled "Seawater Steam Engine as a 'prime mover' for the third industrial revolution", which technology is a scientific and technological breakthrough for stopping climate change.

Dr. Anamarija Farkaš (co-editor) is a Senior Research Associate who graduated in from the Faculty of Chemical Engineering and Technology, University of Zagreb, Croatia, in 1990. Since 1994, she has been working as researcher at the IRMO in the Department for Resource Economics, Environmental Protection and Regional Development. In 2004 she obtained a PhD from the University of Zagreb. Dr. Farkaš' activities are focused on chemistry and ecology, environmental policy, environmental engineering, environmental economy, and bioeconomy. She participated in a project investigating the availability and ecological advisability of applying natural resources in ecological agriculture, forestry protection, livestock, wastewater treatment, and air pollution.

Contents

Preface

"There are no things that would be so valuable teaching as nature"
Nikola Tesla

Today's global challenges such as climate change, pandemics, lack of drinking water, and so on, require scientists and experts to work even harder in finding answers and solutions to such challenges, all with the goal to protect the environment, achieving sustainable development as well as survival of mankind on Earth.

Natural resources, such as zeolite minerals, have an inexhaustible potential for scientific research and application. Based on the research of natural zeolites, synthetic zeolites have been obtained in laboratory conditions, and both types have application in many researched areas (e.g., water, soil, industries, biochemistry, medicine, etc.). Scientific research is intense in still insufficiently researched areas of their application, such as the absorption of greenhouse gases and application of zeolite membranes in virology.

From the discovery of zeolites until today, scientific interest in their research and application has not decreased because zeolites can have a significant role in numerous sustainable processes due to their ecological and economical suitability and especially due to their unique structure and specific physical and chemical properties.

This book is organized into three sections: "Adsorption and Diffusion Processes in Zeolite Materials," "Layered Zeolites," and "Application of Zeolite Materials."

The introductory chapter, "The Past, Present and the Future of Zeolite Science," presents a historic overview of the discovery of zeolite materials. From the literature we find that the basis of mineralogy was set in 371–297 BC (Teofrast). A sudden increase in mineral research was noted in the eighteenth century, and based on Cronstedt's observations the first properties of zeolites were discovered. Further research during the nineteenth century, and in parallel with the development of instrumental techniques, led to the discovery of new properties of zeolite materials, which contributed to intensive growth and production of synthetic zeolites during the twentieth and twenty-first centuries. A great number of scientific research, patents, and industrial applications of zeolites from the twentieth century until today enabled the growth of the global zeolite market. Climate changes, threats to mankind such as the COVID-19 pandemic, and changes in economic and political direction present new challenges for scientists and experts in research and application of these amazing materials.

The first section of the book addresses the understanding of the structure, characteristics, and physical-chemical properties of zeolites as well as diffusion and adsorption processes in research of zeolite materials. The second chapter in this section, "Competitive Adsorption and Diffusion of Gases in a Microporous Solid," develops a new, fast, and accurate analytical method for the calculation of the coefficients of co-diffusing gases in the intra- and inter-crystallite spaces of microporous solid (i.e., ZSM 5 zeolite) using high-performance methods (iterative gradient methods of residual functional minimization and analytical methods of influence functions), mathematical co-adsorption models, and NMR spectra of each adsorbed gas in the bed.

The third chapter in this section, "Zeolites as Chameleon Biomaterials: Adsorption of Proteins, Enzymes, Foods, Drugs, Human Cells, and Metals on Zeolite Membranes with Versatile Physics-Chemical Properties," describes the physical-chemical properties of zeolites (obtained by hydrothermal synthesis) in interaction with transition metals, proteins, enzymes, drugs, food, and human cells as well as the development of physicochemical properties of zeolite membranes for biomedical and biotechnological applications.

The fourth chapter in this section, "Anomalous Diffusivity in Porous Solids: Levitation Effect" presents interesting and extensive research that describes deviations from standard or expected diffusion in porous solids, called the levitation effect. In addition to the simulation of anomalous diffusion in zeolite pores by guest monatomic molecules, simulation is performed on anisotropic molecules (n-pentane, isopentane, and neopentane).

The second section of the book contains one chapter, "New Trends in Layered Zeolites," which examines a new family of microporous materials called layered zeolites. It presents a very interesting overview of recently developed synthetic methods and modifications of layered zeolites as well as possible future development of layered zeolites that could contribute to the expansion of the layered zeolite family.

The third section of the book contains two chapters that discuss scientific research and development of zeolite materials for application in different areas. The first chapter in this section, "Zeolites: An Emerging Material for Gas Storage and Separation Applications," describes the development of zeolite membranes of three different types of zeolite material (DDR, SAPO 34, BIK) for separation of different gas mixtures and storage of CO_2. These zeolites/surface-modified zeolites and their membranes have potential application for environmental protection in the near future. The second chapter in this section, "Zeolites Applications in Veterinary Medicine," provides an extensive overview of zeolite applications in veterinary medicine.

Karmen Margeta
University of Zagreb,
Zagreb, Croatia

Anamarija Farkaš
Institute for Development and International Relations,
Zagreb, Croatia

Section 1

Adsorption and Diffusion Processes in Zeolite Materials

Chapter 1

Introductory Chapter: Zeolites - From Discovery to New Applications on the Global Market

Karmen Margeta and Anamarija Farkaš

1. Introduction

From the discovery of zeolites until today, the interest in scientific research and use of this amazing, interesting, and useful material has continuously grown.

Natural zeolites are hydrated aluminosilicate materials. Primary building units (PBUs) are tetrahedra of Si and Al oxides which are interconnected by oxygen ions into two-dimensional and three-dimensional secondary units (secondary building units [SBUs]). The zeolite network consists of channels and cavities filled with hydrated alkali and alkaline earth metal ions. With the presence of these metal ions in the structure of zeolites, a balance between positive electric charge of metal ions and negative charge of zeolite network is achieved.

The hydrating and dehydrating properties of zeolites were crucial in discovering their physical and chemical properties and their wider use, and that is why this mineral was named "boiling stone" (ζέω (zéō) meaning "to boil" and λίθος (líthos) meaning "stone") [1].

Unlike natural zeolites which are formed as crystals in cavities of basalt rocks or as volcanic tuff in different geological environments at relatively low temperatures, synthetic zeolites are created by hydrothermal synthesis in laboratory conditions. During hydrothermal synthesis in the presence of certain chemical compounds ("template" or structure directing agent (SDA)), which act on the organization of SBUs and consequential formation of tertiary or composite building units (TBUs), a final crystal zeolite is created in the form of a polyhedron which contains Si-O-Al bonds. Control of kinetic processes during zeolite synthesis is especially important since most of the zeolites created via this process are metastable phases [2].

2. Zeolite's history

Zeolite materials were discovered in the eighteenth century when a Swedish chemist, founder of modern mineralogy, Baron Axel Fredrik Cronstedt, discovered that during the heating of mineral "stilbite," moisture appears on its surface [3]. His discovery began the research of chemical, physical, and mineralogical characteristics of natural zeolites.

In spite of the limited capabilities of structural research of zeolites at the time of their discovery (unlike today's modern methods), researchers set the foundations for today's application of natural and synthetic zeolites (**Figure 1**). Based on Cronstedt's observations and research, in the middle of the nineteenth century according to the available literature, the first synthetic zeolite-levinite [5] was synthesized, while in the first half of the twentieth century (after the first structural analysis

IntechOpen

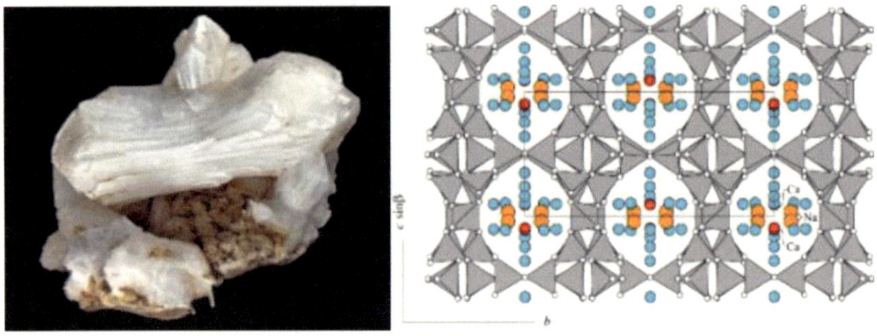

Figure 1.
The mineral stilbite (left) and the crystal structure (framework) of the stilbite group (right) [4].

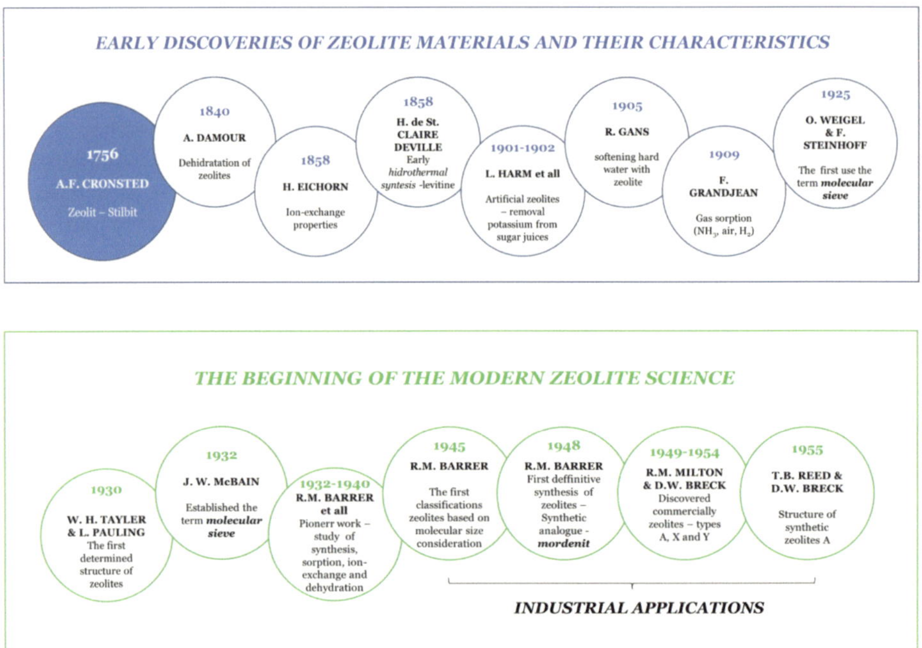

Figure 2.
Early discoveries of zeolite materials and their characteristics, the beginning of the modern zeolite science, and their industrial applications [7, 8].

of zeolite materials), a synthetic analog zeolite-mordenite [6] was synthesized. Intensive research on the synthesis as well as physical and chemical properties of zeolites (absorption, ion exchange, dehydration) enabled the industrial application of zeolites. Significant discoveries and research of zeolites, as well as the beginning of modern zeolite science (parallel with the development of new instrumental techniques), are shown in **Figure 2**.

3. Scientific research and industrial applications of zeolites

Almost three centuries have passed since the discovery of zeolites, and the interest in research of natural zeolite materials has not diminished. Equally, obtaining the optimal properties of synthetic zeolite for their technological application is a priority in scientific research.

In 2007 the sixth edition of "The Atlas of Zeolites" was published, in which the data for each of the 176 unique zeolite framework types are presented [9]. Of the total natural zeolites discovered (more then 60), only six occur in large quantities in the natural deposits worldwide: analcime (ANA), chabazite (CHA), clinoptilolite (HEU), erionite (ERI), mordenite (MOR), and phillipsite (PHI). Ferrierite (FER) occurs in a few large deposits.

Each of the seven natural zeolites has been synthesized, but only mordenite and ferrierite are synthesized in large quantities (synthetic mordenite has large pores while natural mordenite has small pores) [10].

Commercial natural zeolite deposits are primarily found in the USA (volcanic tuffs in saline, alkaline lake deposits, and open hydrologic systems). If both natural and synthetic zeolites are equally available in commercial quantity, synthetic zeolite will be more acceptable because the process of removing impurities from natural zeolite is significantly more expensive.

Even though there is a large number of scientific research which describes properties, structures, characteristics, physical and chemical processes of absorption, and ionic exchange of zeolites with the use of modern techniques, still some of the properties and processes are not completely understood, especially the ones related to natural zeolites. Still there is room for further research and application of these incredible microporous materials in areas which are still unexplored.

Zeolites have found application in almost all areas of agriculture and animal husbandry, biotechnology, medicine, chemical industry, construction industry, oil and natural gas industry, water processing, etc., as well as new technologies which contribute to sustainable development. Due to their specific properties, high thermal and hydrothermal stability, as well as environmental acceptability, zeolites are an alternative to similar materials that are not economically and environmentally acceptable [11–34].

Since the first industrial applications (during the beginning of the twentieth century) until today, demand for zeolites has permanently grown, whether they are used individually or in combination with other micro- and nanoporous materials. New scientific research of synthetic zeolites contributes to the further development of highly active and selective synthetic zeolites used as catalysts, adsorbents, and ion exchangers with high capacity and selectivity.

The increasing number of national and world technological innovation (patents) allows the new industrial applications of zeolite materials, which positively affect economic growth and development with the emphasis on zeolite materials that do not pollute the environment [35, 36].

3.1 Global zeolite market

Zeolite markets (natural and synthetic products) are developed on all continents, shown in **Figure 3**. The largest market of zeolites is in the USA, followed by the Asia-Pacific area, China, India, Europe, and other markets.

Global zeolite market share (by products) shows that the synthetic zeolites have a bigger share than the natural zeolites (**Figure 4**).

In 2016 the zeolite global market was estimated at 29.08 billion USD with an expected growth of 2.5% in the period from 2016 to 2022. Therefore, zeolite global market could achieve a value of 33.8 USD until 2022 [37].

Market research companies have published reports in which the following are presented: zeolite market size, share, and trends analysis report by product, by application, and by region, for the specific estimated period. Some of their data are shown in **Table 1**. The value of compounded annual rate of growth (CAGR) is referring to the valuation before the COVID-19 pandemic.

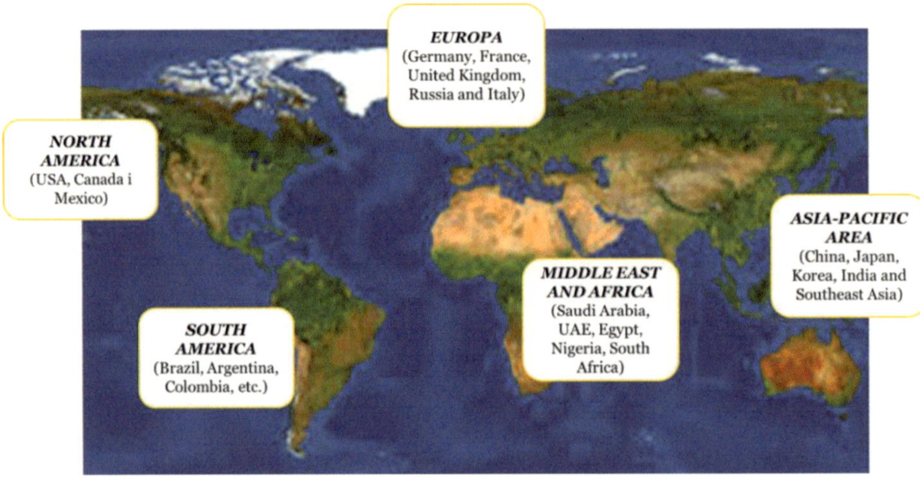

Figure 3.
Zeolite market segmented by continents.

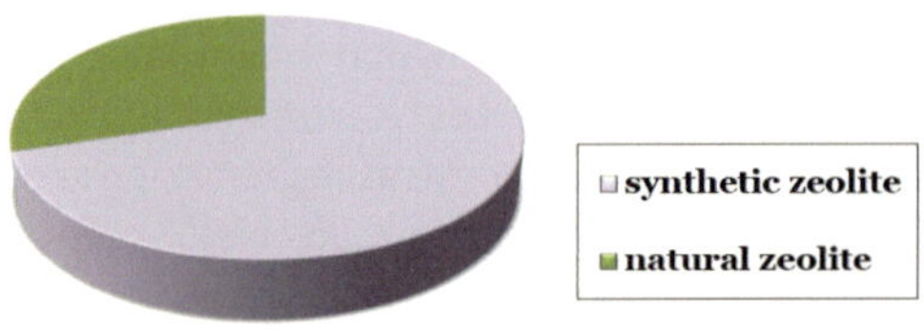

Figure 4.
Global zeolite market share (by products).

Natural and synthetic zeolite/application	Zeolite market size	Forecast period	CAGR	Ref. (*MRC)
High-quality transportation fuels and chemical products	4.15 billion USD	2019–2024	5.3%	[39]
Membrane	**NA	2017–2025	3.7%	[40]
Paints and coating industry	53.4 million USD	2016–2024	3.2%	[41]
Detergents	1.4 billion USD	2018–2028	2.6%	[42]
Adsorbents, ion exchange, catalyst	**NA	2019–2026	2.73%	[43]
Cement, animal feed industry, wastewater treatment	6.6 billion USD	2019–2026	4.9%	[44]
Synthetic zeolite market	19.81 billion USD	2017–2025	4.6%	[45]

**MRC, market research company.*
***NA, not available.*

Table 1.
Estimation of global zeolite markets growth (data were taken from market research company reports).

Estimations of market research companies are that the market of natural zeolites will have slower growth until the year 2030, considering limitations such as the high price of raw materials, high costs of transportation of raw materials, and availability of alternative materials in specific areas of industrial application (such as enzymes, metals, and other chemical compounds in catalysis processes). However, natural zeolites hold an advantage over other materials because they are nontoxic and ecologically acceptable and as such materials can meet the demands which are set by environment protection laws.

Unlike natural zeolites, the demand for synthetic zeolites until the year 2030 is estimated to grow in areas of industrial application [38].

Global markets depend on a series of factors such as political and economic situations as well as other factors that cannot be influenced or predicted. One such factor is the COVID-19 pandemic which (according to first forecasts) caused a change in market conditions and slowed down technological progress and the start of new production, as well as made launching new products on the market based on zeolite materials harder.

4. New challenges in research and application of zeolites

Today, the humankind is faced with many threats such as (i) drought and lack of freshwater caused by climate change, (ii) political and economic instabilities, and (iii) dangerous pandemics such as the SARS-CoV-2 pandemic, which caused slowing down of scientific research, economic growth, and development and even brought into question the survival of mankind.

With current use of zeolites in agriculture, animal husbandry, water processing, numerous industries (chemical, oil, construction, etc.) biotechnology, medicine, etc., the future of zeolite application opens up new possibilities for research, development, and technological application in still undiscovered areas such as medication and antimicrobial application [46], which would allow new global markets for zeolites.

Furthermore, the application of zeolites in technologies that can contribute to the reduction of CO2 emissions (one of the greenhouse gases which contributes to climate change) could have a significant role in their future use [47–49].

5. Conclusions

Scientific research of natural zeolites poses a challenge in zeolite science because there are still many unexplored properties of this fascinating, ecologically friendly, and truly useful microporous material. Further research on natural zeolite and production of new synthetic zeolites open up possibilities of their application in still unexplored areas in the near future because their application requires low technical complexity and initial infrastructure investment.

Conflict of interest

The authors declare no conflict of interest.

Author details

Karmen Margeta[1*] and Anamarija Farkaš[2]

1 University of Zagreb, Zagreb, Croatia

2 Institute for Development and International Relations, Zagreb, Croatia

*Address all correspondence to: karmen.margeta@gmail.com

References

[1] Breck DW. Zeolite Molecular Sieves: Structure, Chemistry, and Use. New York: Wiley; 1973

[2] Berenguer MÁ. Ordered porous nanomaterials: The merit of small. International Scholarly Research Notices. 2013:1-29. DOI: 10.1155/2013/257047

[3] Cronstedt AF. Natural zeolite and minerals. Svenska Vetenskaps Akademiens Handlingar Stockholm. 1756;**17**:120-123

[4] IZA Commission on Natural Zeolite. 2005. Available from: http://www.iza-online.org/natural/Datasheets/Stilbite/stilbite.htm

[5] Annales des mines. 1999. Available from: http://annales.ensmp.fr/numeros/ANN_1865_S06_08/ANN_1865_S06_08.pdf

[6] Pramod K, Bajpai M, Rao S, Gokhale KVGK. Synthesis of mordenite type zeolites. Industrial and Engineering Chemistry Product Research and Development. 1978;**17**(3):223-227. DOI. 10.1021/i360067a009

[7] Heaney PJ, Prewitt CT, Gibbs GV, editors. Silica: Physical behavior, geochemistry, and materials applications. Reviews in Mineralogy. Mineralogical Society of America. 1994;**29**(1):1-40

[8] Jacobs PA, Flanigen EM, Jansen JC, Bekkum H, editors. Introduction to Zeolite Science and Practice. 2nd Completely Revised and Expanded Edition. Amsterdam, New York: Elsevier; 2001. p. 1078

[9] Baerlocher C, McCusker LB, Olsen DH. Atlas of Zeolite Framework Types. 6th ed. Amsterdam: Elsevier; 2007. p. 398

[10] John D, Sherman JD. Synthetic zeolites and other microporous oxide molecular sieves. Proceedings of the National Academy of Science of the USA. 1999;**96**(7):3471-3478. DOI: 10.1073/pnas.96.7.3471. Washington. Available from: https://www.pnas.org/content/pnas/96/7/3471.full.pdf

[11] Zimmermann NER, Haranczyk M. History and utility of zeolite framework-type discovery from aData-science perspective. Crystal Growth & Design. 2016;**16**:3043-3048. DOI: 10.1021/acs.cgd.6b00272

[12] Bacakova L, Vandrovcova M, Kopovaaand I, Jirka I. Applications of zeolites in biotechnology and medicine–a review. Biomaterials Science. 2018;**6**(5):974-989. DOI: 10.1039/c8bm00028j

[13] Montalvo S, Huiliñir C, Borja R, Sánchez E, Herrmann C. Application of zeolites for biological treatment processes of solid wastes and wastewaters—A review. Bioresource Technology. 2020;**301**. DOI: 10.1016/j.biortech.2020.122808

[14] Derakhshankhah H, Jafari S, Sarvari S, Barzegari E, Moakedi F, Ghorbani M, et al. Biomedical applications of Zeolitic nanoparticles, with an emphasis on medical interventions. International Journal of Nanomedicine. 2020;15:363-386. DOI: 10.2147/IJN.S234573

[15] Li Y, Li L, Yu J. Applications of zeolites in sustainable chemistry (a review). Chem. 2017;**3**(6):928-949. DOI: 10.1016/j.chempr.2017.10.009

[16] Moshoeshoe M, Nadiye-Tabbiruka MS, Obuseng V. A review of the chemistry, structure, properties and applications of zeolites American. Journal of Materials Science. 2017;**7**(5):196-221

[17] Kraljević Pavelić S, Simović Medica J, Gumbarević D, Filošević A,

Pržulj N, Pavelić K. Critical review on zeolite clinoptilolite safety and medical applications in vivo. Frontiers in Pharmacology. 2018;**9**:1350. DOI: 10.3389/fphar.2018.01350

[18] Gläser R, Weitkamp J. The application of zeolites in catalysis. Basic principles in applied catalysis. In: Baerns M, editor. Springer Series in Chemical Physics. Vol. 75. Berlin, Heidelberg: Springer; 2004. p. 211. DOI: 10.1007/978-3-662-05981-4_5

[19] Nakhli SAA, Delkash M, Bakhshayesh BE, et al. Application of zeolites for sustainable agriculture: A review on water and nutrient retention. Water, Air, and Soil Pollution. 2017;**228**:464. DOI: 10.1007/s11270-017-3649-1

[20] Woszuk A, Franus W. A review of the application of zeolite materials in warm mix asphalt technologies. Applied Sciences. 2017;**7**(3):293. DOI: 10.3390/app7030293

[21] Wang S, Peng Y. Natural zeolites as effective adsorbents in water and wastewater treatment. Chemical Engineering Journal. 2010;**156**:11-24

[22] Margeta K, Zabukovec Logar N, Šiljeg M, Farkaš A. Natural zeolites in water treatment—How effective is their use. In: Elshorbagy W, Rezaul Kabir Chowdhury RK, editors. Water Treatment. Rijeka, London: InTechOpen; 2013. pp. 82-112. DOI: 10.5772/50738

[23] Dyer A, Keir D. Nuclear waste treatment by zeolites. Zeolites. 1984;**4**(3):215-217. DOI: 10.1016/0144-2449(84)90026-5

[24] Lee HY, Kim HS, Jeong H-K, Park M, Chung D-Y, Lee K-Y, et al. Selective removal of radioactive cesium from nuclear waste by zeolites: On the origin of cesium selectivity revealed by systematic crystallographic studies. The Journal of Physical Chemistry C. 2017;**121**(19):10594-10608. DOI: 10.1021/acs.jpcc.7b02432

[25] Ghasemi Z, Sourinejad I, Kazemian H, Rohani S. Application of zeolites in aquaculture industry: A review. Reviews in Aquaculture. 2014;**10**(1):75-95. DOI: 10.1111/raq.12148

[26] Williams CD. Application of zeolites to environmental remediation. In: Charlesworth SM, Booth CA, editors. Urban Pollution: Science and Management. New York: Wiley; 2018. DOI: 10.1002/9781119260493.ch19

[27] Singh MP, Baghel GS, Titinchmi SJJ, Abbo HS. Zeolites: Smart materials for novel, efficient, and versatile catalysis. In: Ashutosh Tiwari A, Titinchi S, editors. Advanced Catalytic Materials. 2015. DOI: 10.1002/9781118998939.ch11

[28] Mravec D, Hudec J, Janotka I. Some possibilities of catalytic and noncatalytic utilization of zeolites. Chemical Papers. 2005;**59**(1):62-69. DOI: 10.1002/chin.200534287

[29] Ackley MW, Rege SU, Saxena H. Review-application of natural zeolites in the purification and separation of gases. Microporous and Mesoporous Materials. 2003;**61**:25-42

[30] Bright KR, Sicairos-Ruelas EE, Gundy PM, Gerba CP. Assessment of the antiviral properties of zeolites containing metal ions. Food and Environmental Virology. 2009;**1**(1):37. DOI: 10.1007/s12560-008-9006-1

[31] Derakhshankhah H, Jafari S, Sarvari S, Barzegari E, Moakedi F, Ghorbani M, et al. Biomedical applications of zeolitic nanoparticles, with an emphasis on medical interventions. International Journal of Nanomedicine. 2020;**15**:363-386. DOI: 10.2147/IJN.S234573

[32] Park DH, Joe YH, Hwang J. Dry aerosol coating of anti-viral particles on commercial air filters using a high-volume flow atomizer. Aerosol and Air Quality Research. 2019;**19**:1636-1644 https://doi.org/10.4209/aaqr.2019.04.0212

[33] Ferreira L, Guedes JF, Almeida-Aguiar C, Fonseca AM, Neves IC. Microbial growth inhibition caused by Zn/Ag-Y zeolite materials with different amounts of silver. Colloids and Surfaces. B, Biointerfaces. 2016;**142**:141-147. DOI: 10.1016/j.colsurfb.2016.02.042

[34] Savi GD, Cardoso WA, Furtado BG, Bortolotto T, Da Agostin LOV, Nones J, et al. New ion-exchanged zeolite derivatives: Antifungal and antimycotoxin properties against *Aspergillus flavus* and aflatoxin B1. Materials Research Express. 2017;**4**:8. Available from: https://iopscience.iop.org/article/10.1088/2053-1591/aa84a5/meta

[35] Margeta K, Vojnović B, Zabukovec Logar N. Development of natural zeolites for their use in water-treatment systems. Recent Patents of Nanotechnology. 2011;**5**(2):89-99. DOI: 10.2174/187221011795909170

[36] Li B, Duan Y, Luebke D, Morreale B. Advances in CO2 capture technology: A patent review. Applied Energy. 2013;**102**:1439-1447

[37] Grand View Research. 2018. Available from: https://www.grandviewresearch.com/industry-analysis/zeolites-market

[38] Fact MR. 2020. Available from: https://www.factmr.com/report/3817/zeolites-market

[39] Market research. 2019. Available from: https://www.marketresearch.com/IMARC-v3797/Zeolite-Global-Trends-Share-Size-12454492/

[40] Credence Research. 2017. Available from: https://www.credenceresearch.com/press/global-zeolites-market

[41] Transparency Market Research. 2016. Available from: https://www.transparencymarketresearch.com/zeolites-market.html

[42] Future Market Insights. 2018. Available from: https://www.futuremarketinsights.com/reports/zeolite-for-detergents-market

[43] Verified Market Research. 2018. Available from: https://www.verifiedmarketresearch.com/product/zeolite-powder-market/

[44] Fortune Business Insights. 2018. Available from: https://www.fortunebusinessinsights.com/industry-reports/refinery-catalyst-market-101090

[45] Global information. 2017. Available from: https://www.giiresearch.com/report/accu530047-global-synthetic-zeolites-market-analysis-trends.html

[46] Nakamura S, Sato M, Sato Y, Ando N, Takayama T, Fujita M, et al. Synthesis and application of silver nanoparticles (Ag NPs) for the prevention of infection in healthcare workers. International Journal of Molecular Sciences. 2019;**20**(15):3620. DOI: 10.3390/ijms20153620

[47] Chouikhi N, Cecilia JA, Vilarrasa-García E, Besghaier S, Chlendi M, Duro FIF, et al. CO_2 adsorption of materials synthesized from clay minerals: A review. Minerals. 2019;**9**:514. DOI: 10.3390/min9090514

[48] Oschatz M, Antonietti M. A search for selectivity to enable CO2 capture with porous adsorbents. Energy &

Environmental Science. 2018;**11**:57. DOI: 10.1039/c7ee02110k

[49] Younas M, Sohail M, Leong LK, Bashir MJK, Sumathi S. Feasibility of CO2 adsorption by solid adsorbents: a review on low-temperature systems. International journal of Environmental Science and Technology. 2016;**13**:1839-1860. DOI: 10.1007/s13762-016-1041-0

Chapter 2

Competitive Adsorption and Diffusion of Gases in a Microporous Solid

Mykhaylo Petryk, Mykola Ivanchov, Sebastian Leclerc, Daniel Canet and Jacques Fraissard

Abstract

The experimental and theoretical study of the co-adsorption and co-diffusion of several gases through a microporous solid and the instantaneous (out of equilibrium) distribution of the adsorbed phases is particularly important in many fields, such as gas separation, heterogeneous catalysis, purification of confined atmospheres, reduction of exhaust emissions contributing to global warming, etc. The original NMR imaging technique used gives a signal characteristic of each adsorbed gas at each instant and at each level of the solid and therefore the distribution of several gases in competitive diffusion and adsorption. But it does not allow to determine separately the inter- and intra-crystallite quantities. A new fast and accurate analytical method for the calculation of the coefficients of co-diffusing gases in the intra- and inter-crystallite spaces of microporous solid (here ZSM 5 zeolite) is developed, using high-performance methods (iterative gradient methods of residual functional minimization and analytical methods of influence functions) and mathematical co-adsorption models, as well as the NMR spectra of each adsorbed gas in the bed. These diffusion coefficients and the gas concentrations in the inter- and intra-crystallite spaces are obtained for each position in the bed and for different adsorption times.

Keywords: competitive diffusion of gases, competitive adsorption, modeling, diffusion coefficient, Heaviside's operational method, zeolite bed, gradient methods of identification

1. Introduction

Knowledge of the co-diffusion and co-adsorption coefficients of reactants and products is essential when a heterogeneous catalytic reaction is performed by flowing gases through a microporous catalyst bed. But generally the distribution of the various reactants adsorbed on the catalyst is very inhomogeneous and, moreover, very variable from one reactant to another. It is therefore necessary to determine at every moment the diffusion coefficient of each reactant in the presence of the others and its instantaneous distribution along the length of the catalyst bed.

Classical H-MRI should be a good technique for monitoring the co-diffusion and co-adsorption of several gases flowing through a microporous bed. However, since the signal obtained is not specific for each gas, this requires that each experiment be performed several times under identical conditions and each time with only one not deuterated gas. To remedy the drawbacks of classical imaging, we have used the NMR imaging technique, named slice selection procedure, to follow the diffusion and adsorption of a gas in a microporous bed [1–3]. The sample is displaced vertically, step-by-step, relative to a very thin coil detector during the adsorption of the gas. The bed is assumed to consist of N very thin layers of solid, and the region probed is limited to each layer, so that the variation of the concentration of gas absorbed at the level of each layer is obtained as a function of time. An interesting feature of this technique is its ability to visualize directly the co-diffusion of several gases. Indeed, the NMR signals are quantitatively characteristic of the adsorbed gases. They can therefore provide directly, at every moment and at every level of the bed, the distribution of several gases competing in diffusion and adsorption. We have presented in a previous paper the experimental results of the co-diffusion of benzene and hexane through a silicalite bed [4]. In [5, 6] we have developed a mathematical methodology for efficient linearization of similar models. Using Heaviside's operational method and Laplace's integral transformation method, we have built solutions allowing fast calculations for two-component co-adsorption in a heterogeneous zeolite bed and for the dehydration of natural gas [7]. In this chapter we have improved the methods previously used to compute the diffusion coefficients against time, increasing the accuracy and speed of calculations by significantly reducing the iteration number. This made it possible to use them for the co-adsorption of several gases diffusing along such a column.

2. Experimental

The NMR imaging technique, the sample-holder bulb containing the liquid phase in equilibrium with the gas phase, and the narrow zone monitored by the detector have been described in [1–3], respectively.

The upper face of the cylindrical bed of zeolite crystallites is exposed to a constant pressure of each gas (**Figure 1**). The diffusion of the two gases is axial in

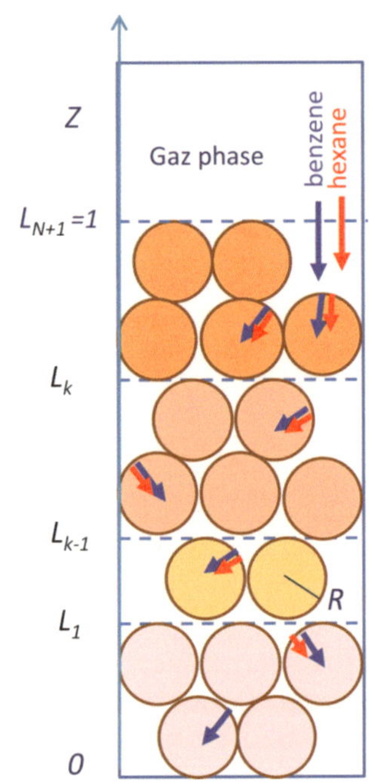

Figure 1.
Distribution of the layers (left) and corresponding parameters (right).

the macropores of the inter-crystallite space (z direction along the height, l, of the bed) and radial in the micropores of the zeolite. According to the experimental conditions, the zeolite bed consists of a large number, N, of very thin layers of solid, of thickness $\Delta l_k = l_k - l_{k-1}$, perpendicular to the propagation of the gas in the z direction. The corresponding coefficients of inter- and intra-crystallite space are $D_{inter,k}$ and $D_{intra,k}$, respectively.

3. Experimental results: gaseous benzene and hexane adsorption curves

The experimental results have been summarized in [4]: the spectrum of each gas at every instant and every level of the solid and the benzene and hexane concentrations along the sample, for each diffusion time. Here we shall only use the evolution, as a function of time, of the benzene and hexane concentrations at different levels of the sample, on which the calculations of the diffusion coefficients and the instantaneous inter- and intra-crystallite concentrations are based [8]. **Figure 2** shows clearly that, under the chosen experimental conditions, benzene hinders the diffusion of hexane, and this at every moment. Moreover, it can be noticed that, at equilibrium, the amount of benzene within the zeolite is twice that of hexane, indicating quantitatively the relative affinity to the two adsorbates.

These curves display modulations as a function of time, which must be averaged for all subsequent mathematical representations. These modulations are weak at the lower layers of the tube and can be due to errors in the measurement of small amounts. Those closer to the arrival of the gas are greater and are similar for the two gases. We suggested that these fluctuations may be due to the fact that inter-crystallite adsorption at levels close to the gas phase is fast compared to the liquid-gas equilibrium, which is not as instantaneous for a mixture as for a single component [8]. Each slight decrease of the gas pressure could correspond a slight fast desorption.

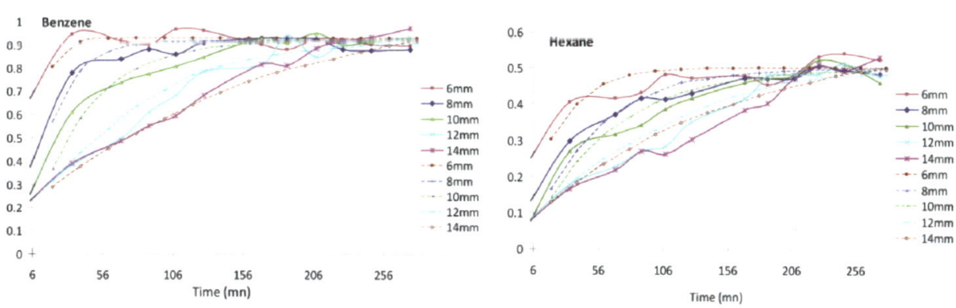

Figure 2.
*Evolution vs. time of the benzene and hexane concentrations (arbitrary units) at different levels of the sample (*continuous, *experimental curves;* dotted, *their approximations used for simulation) [from Ref. [8], reprinted with permission from ACS].*

4. A mathematical model of competitive co-adsorption and co-diffusion in microporous solids

4.1 Co-adsorption model in general formulation

The model presented is similar to the bipolar model [2, 3, 8, 9]. By developing the approach described by Ruthven and Kärger [10, 11] and Petryk et al. [5]

concerning the elaboration of a complex process of co-adsorption and co-diffusion, it is necessary to specify the most important hypotheses limiting the process.

The general hypothesis adopted to develop the model presented in the most general formulation is that the interaction between the co-adsorbed molecules of several gases and the adsorption centers on the surface in the nanoporous crystallites is determined by the nonlinear competitive equilibrium function of the Langmuir type, taking into account physical assumptions [10]:

1. Co-adsorption is caused by the dispersion forces whose interaction is established by Lennard-Jones and the electrostatic forces of gravity and repulsion described by van der Waals [11].

2. The co-diffusion process involves two types of mass transfer: diffusion in the macropores (inter-crystallite space) and diffusion in the micropores of crystallites (intra-crystallite space).

3. During the evolution of the system toward equilibrium, there is a concentration gradient in the macropores and/or in the micropores.

4. Co-adsorption occurs on active centers distributed over the entire inner surface of the nanopores (intra-crystallite space) [10, 11]. All crystallites are spherical and have the same radius R; the crystallite bed is uniformly packed.

5. Active adsorption centers adsorb molecules of the ith adsorbate, forming molecular layers of adsorbate on their surfaces.

6. Adsorbed molecules are held by active centers for a certain time, depending on the temperature of the process.

Taking into account these hypotheses, we have developed a nonlinear co-adsorption model. The meaning of the symbols is given in the nomenclature:

$$\frac{\partial C_s(t,Z)}{\partial t} = \frac{D_{\text{inter}_s}}{l^2}\frac{\partial^2 C_s}{\partial Z^2} - e_{\text{inter}}\tilde{K}_s\frac{D_{\text{intra}_s}}{R^2}\left(\frac{\partial Q_s}{\partial X}\right)_{X=1} \quad (1)$$

$$-H\frac{\partial T(t,z)}{\partial t} - uh_g\frac{\partial T}{\partial z} - \sum_{s=1}^{m}\Delta\overline{H}_s\frac{\partial\overline{Q}_s}{\partial t} - 2\frac{\alpha_h}{R_{column}}T + \Lambda\frac{\partial^2 T}{\partial z^2} = 0 \quad (2)$$

$$\frac{\partial Q_s(t,X,Z)}{\partial t} = \frac{D_{\text{intra}_s}}{R^2}\left(\frac{\partial^2 Q_s}{\partial X^2} + \frac{2}{X}\frac{\partial Q_s}{\partial X}\right) \quad (3)$$

with initial conditions

$$C_s(t=0,Z) = 0; Q_s(t=0,X,Z) = 0; Z \in (0,1), X \in (0,1), s = \overline{1,\mathrm{m}} \quad (4)$$

boundary conditions for coordinate X of the crystallite

$$\frac{\partial}{\partial X}Q_s(t,X=0,Z) = 0 \text{ (symmetry conditions)}, \quad (5)$$

$$Q_s(t,X=1,Z) = \frac{K_s(T)C_s(t,Z)}{1+\sum_{s_1=1}^{m}K_{s_1}(T)C_{s_1}(t,Z)}, s = \overline{1,\mathrm{m}} \text{ (Langmuir equilibrium)}, \quad (6)$$

boundary and interface conditions for coordinate Z

$$C_s(t,1) = 1, \frac{\partial C_s}{\partial Z}(t, Z=0) = 0, t>0 \tag{7}$$

$$T(t,Z)\Big|_{Z=1} = T_{initial}, \frac{\partial}{\partial z}T(t,Z)|_{Z=0} = 0. \tag{8}$$

with $K_s(T) = k_{0s}\exp\left(-\frac{\Delta H_s}{R_g T}\right)$.

Here the activation energy is the heat of adsorption defined as $\Delta H_s = \overline{\phi} - \left(U_{g_s} - U_{ads_s}\right) - R_g T$, where $U_{g_s} - U_{ads_s}$—the difference between the kinetic energies of the molecule of the ith component of the adsorbate in the gaseous and adsorbed states is the magnitude of the Lennard-Jones potential, averaged over the pore volume of the adsorbent [11].

The non-isothermal model (1)–(8) can easily be transformed into isothermal model, removing the temperature Eq. (2) and condition (8) and replacing the functions $K_s(T)$ with the corresponding equilibrium constants K_s. The competitive diffusion coefficients D_{intra_s} and D_{inter_s} can be considered as functions of the time and the position of the particle in the zeolite bed.

4.2 The inverse model of co-diffusion coefficient identification: application to the benzene-hexane mixture

On the basis of a developed nonlinear co-adsorption model (1)–(8), we construct an inverse model for the identification of the competitive diffusion coefficients D_{intra_s} and D_{inter_s} as a function of time and coordinate in the zeolite bed.

The mathematical model of gas diffusion kinetics in the zeolite bed is defined in domains: $\Omega_{k_t} = \left(0, t^{total}\right) \times \Omega_k$, $\left(\Omega_k = (L_{k-1}, L_k), k = \overline{1, N+1}, L_0 = 0 < L_1 < \ldots < L_{N+1} = 1\right)$ by the solutions of the system of differential equations

$$\frac{\partial C_{s_k}(t,Z)}{\partial t} = \frac{D_{\text{inter}_{s_k}}}{l^2}\frac{\partial^2 C_{s_k}}{\partial Z^2} - e_{\text{inter}_k}\tilde{K}_{s_k}\frac{D_{\text{intra}_{s_k}}}{R^2}\left(\frac{\partial Q_{s_k}}{\partial X}\right)_{X=1} \tag{9}$$

$$\frac{\partial Q_{s_k}(t,X,Z)}{\partial t} = \frac{D_{\text{intra}_{s_k}}}{R^2}\left(\frac{\partial^2 Q_{s_k}}{\partial X^2} + \frac{2}{X}\frac{\partial Q_{s_k}}{\partial X}\right) \tag{10}$$

with initial conditions

$$C_{s_k}(t=0,Z) = 0; Q_{s_k}(t=0,X,Z) = 0; X \in (0,1), Z \in \Omega_k, k = \overline{1, N+1}, \tag{11}$$

boundary and interface conditions for coordinate Z

$$C_{s_1}(t, L_1) = 1, \frac{\partial C_{s_1}}{\partial Z}(t, Z=0) = 0, t \in \left(0, t^{total}\right); \tag{12}$$

$$\left[C_{s_k}(t,Z) - C_{s_k}(t,Z)\right]_{Z=L_k} = 0, \frac{\partial}{\partial Z}\left[D_{\text{inter}_{s_{k-1}}}C_{s_{k-1}}(t,Z) - D_{\text{inter}_{s_k}}C_{s_k}(t,Z)\right]_{Z=L_k} = 0, k = \overline{1, N}, t \in \left(0, t^{total}\right); \tag{13}$$

boundary conditions for coordinate X in the particle

$$\frac{\partial}{\partial X}Q_{s_k}(t,X=0,Z)=0,Q_{s_k}(t,X=1,Z)$$
$$=K_sC_{s_k}(t,Z)\ (\text{equilibrium conditions}),Z\in\Omega_k,k=\overline{1,N+1}. \tag{14}$$

Additional condition (NMR-*experimental data*)

$$\left[C_{s_k}(t,Z)+\overline{Q}_{s_k}(t,Z)\right]\Big|_{h_k}=M_{s_k}(t,Z)\Big|_{h_k},\ s=\overline{1,2};h_k\in\Omega_k. \tag{15}$$

The problem of the calculation (9)–(15) is to find unknown functions $D_{\text{intra}_s}\in\Omega_t, D_{\text{inter}_s}\in\Omega_t$ ($D_{\text{intra}_s}>0, D_{\text{inter}_s}>0, s=\overline{1,2}$), when absorbed masses $C_{s_k}(t,Z)+\overline{Q}_{s_k}(t,Z)$ satisfy the condition (15) for every point $h_k\subset\Omega_k$ of the kth layer [8, 12].

Here

$$e_{\text{inter}_k}=\frac{\varepsilon_{\text{inter}_k}c_{sk}}{\varepsilon_{\text{inter}_k}c_{sk}+\left(1-\varepsilon_{\text{inter}_k}\right)q_{sk}}\approx\frac{\varepsilon_{\text{inter}_k}}{\left(1-\varepsilon_{\text{inter}_k}\right)\tilde{K}_{sk}},\ \tilde{K}_{s_k}=\frac{q_{s_k\infty}}{c_{s_k\infty}},$$

where $\overline{Q}_s(t,Z)=\int_0^1 Q_s(t,X,Z)XdX$ is the average concentration of adsorbed component s in micropores and $M_s(t,Z)|_{h_k}$ is the experimental distribution of the mass of the sth component absorbed in macro- and micropores at $h_k\subset\Omega_k$ (results of NMR data, **Figure 2**).

4.3 Iterative gradient method of co-diffusion coefficient identification

The calculation of $D_{\text{intra}_{s_k}}$ and $D_{\text{inter}_{s_k}}$ is a complex mathematical problem. In general, it is not possible to obtain a correct formulation of the problem (9)–(15) and to construct a unique analytical solution, because of the complexity of taking into account all the physical parameters (variation of temperature and pressure, crystallite structures, nonlinearity of Langmuir isotherms, etc.), as well as the insufficient number of reliable experimental data, measurement errors, and other factors.

Therefore, according to the principle of Tikhonov and Arsenin [13], later developed by Lions [14] and Sergienko et al. [15], the calculation of diffusion coefficients requires the use of the model for each iteration, by minimizing the difference between the calculated values and the experimental data.

The calculation of the diffusion coefficients (9)–(15) is reduced to the problem of minimizing the functional of error (16) between the model solution and the experimental data, the solution being refined incrementally by means of a special calculation procedure which uses fast high-performance gradient methods [6, 8, 12, 15].

According to [12, 15], and using the error minimization gradient method for the calculation of $D_{\text{intra}_{s_k}}$ and $D_{\text{inter}_{s_k}}$ of the sth diffusing component, we obtain the iteration expression for the n + 1th calculation step:

$$D^{n+1}_{\text{intra}_{s_k}}(t)=D^n_{intra_{s_k}}(t)-\nabla J^n_{D_{\text{intra}_{s_k}}}(t)$$
$$\frac{\left[C_{s_k}\left(D^n_{\text{inter}_{s_k}},D^n_{\text{intra}_{s_k}};t,h_k\right)+\overline{Q}_{s_k}\left(D^n_{\text{inter}_{s_k}},D^n_{\text{intra}_{s_k}};t,h_k\right)-M_{s_k}(t)\right]^2}{\left\|\nabla J^n_{D_{\text{intra}_{s_k}}}(t)\right\|^2+\left\|\nabla J^n_{D_{\text{inter}_{s_k}}}(t)\right\|^2},$$
$$D^{n+1}_{\text{inter}_{s_k}}(t)=D^n_{\text{inter}_{sm}}(t)-\nabla J^n_{D_{\text{inter}_{s_k}}}(t)$$

$$\frac{\left[C_{s_k}\left(D^n_{\text{inter}_{s_k}}, D^n_{\text{intra}_{s_k}}; t, h_k\right) + \overline{Q}_{s_k}\left(D^n_{\text{inter}_{s_k}}, D^n_{\text{intra}_{s_k}}; t, h_k\right) - M_{s_k}(t)\right]^2}{\left\|\nabla J^n_{D_{\text{intra}_{s_k}}}(t)\right\|^2 + \left\|\nabla J^n_{D_{\text{inter}_{sk}}}(t)\right\|^2}, \quad t \in \left(0, t^{total}\right) \tag{16}$$

where $J\left(D_{\text{inter}_{s_k}}, D_{\text{intra}_{s_k}}\right)$ is the error functional, which describes the deviation of the model solution from the experimental data on $h_k \in \Omega_k$, which is written as

$$J\left(D_{\text{inter}_{s_k}}, D_{\text{intra}_{s_k}}\right) = \frac{1}{2}\int_0^T \left[C_s\left(\tau, Z, D_{\text{inter}_{s_k}}, D_{\text{intra}_{s_k}}\right) + \overline{Q}_s\left(t, Z, D_{\text{inter}_{s_k}}, D_{\text{intra}_{s_k}}\right) - M_{s_k}(t)\right]^2_{h_k} d\tau,$$

$$h_k \in \Omega_k, k = \overline{1, N+1}, \tag{17}$$

$\nabla J^n_{D_{\text{intter}_{s_k}}}(t)$, $\nabla J^n_{D_{\text{intra}_{s_k}}}(t)$ (the gradients of the error functional), $J\left(D_{\text{inter}_{s_k}}, D_{\text{intra}_{s_k}}\right)$.

$\left\|\nabla J^n_{D_{\text{inter}_{sk}}}(t)\right\|^2 = \int_0^T \left[\nabla J^n_{D_{\text{inter}_{sk}}}(t)\right]^2 dt$, $\left\|\nabla J^n_{D_{\text{intra}_{sk}}}(t)\right\|^2 = \int_0^T \left[\nabla J^n_{D_{\text{intra}_{sk}}}(t)\right]^2 dt.$

4.4 Analytical method of co-diffusion coefficient identification

With the help of iterative gradient methods on the basis of the minimization of the residual functional, very precise and fast analytical methods have been developed making it possible to express the diffusion coefficients in the form of time-dependent analytic functions (16). For their efficient use, it is necessary to have an extensive experimental database, with at least two experimental observation conditions for the simultaneous calculation of $D_{\text{intra}_{s_k}}$ and $D_{\text{inter}_{s_k}}$ coefficients. Our experimental studies were carried out for five Z positions of the swept zeolite layer for each of the adsorbed components. The data were not sufficient to fully apply this simultaneous identification method to these five sections. We therefore used a combination of the analytical method and the iterative gradient method for determining the co-diffusion coefficients.

Using Eqs. (9)–(15), it is possible to calculate $D_{\text{intra}_{s_k}}, D_{\text{inter}_{sk}}$ as a function of time using the experimental data obtained by NMR scanning. In particular, in Eqs. (9) and (10), the co-diffusion coefficients can be set directly as functions of the time t: $D_{\text{intra}_{s_k}}(t), D_{\text{inter}_{sk}}(t)$. In this case, the boundary condition (11) can be given in a more general form—also as a function of time:

$$C_{s_1}(t, 1) = C_s^{initial}(t). \tag{18}$$

Experimental NMR scanning conditions are defined simultaneously for all P observation surfaces:

$$\left[C_{sk}(t, Z) + \int_0^1 Q_{sk}(t, X, Z)dX\right]_{Z=h_i} = M_{sk_i}(t, Z)\big|_{h_i}, i = \overline{1, P}, s = \overline{1, 2}; h_i \in \bigcup_{k=1}^{N+1} \Omega_k \tag{19}$$

For simplicity we design $u(t, Z) = C_{sk}(t, Z), v(t, \text{X}, Z) = Q_{sk}(t, \text{X}, Z)$,

$$b(t) = D_{\text{intra}_{sk}}(t)/\mathrm{R}^2, \chi_i(t) = M_{sk_i}(t), i = \overline{1, \mathrm{P}},$$

and considering Eq. (10) in flat form, its solution can be written as [16]:

$$v(t, \mathrm{X}, \mathrm{Z}) = -\int_0^t \mathcal{H}_{4\xi}^{(2)}(t, \tau, \mathrm{X}, 1))b(\tau)u(\tau, \mathrm{Z})d\tau \tag{20}$$

where $\mathcal{H}_{4\xi}^{(2)}(t, \tau, \mathrm{X}, \xi) = -2\sum_{m=0}^{\infty} e^{-\eta_m^2(\theta_2(\mathrm{t})-\theta_2(\tau))}\eta_m \cos\eta_m X \cdot (-1)^m$.

Here the Green influence function of the particle $\mathcal{H}_k^{(2)}, \overline{k = 1, 4}$ is used; it has the form [17].

$$\mathcal{H}_4^{(2)}(t, \tau, \mathrm{X}, \xi) = 2\sum_{m=0}^{\infty} e^{-\eta_m^2(\theta_2(\mathrm{t})-\theta_2(\tau))}\cos\eta_m X \cos\eta_m\xi, \eta_m = \frac{2m+1}{2}\pi,$$
$$\mathcal{H}_3^{(2)}(t, \tau, \mathrm{X}, \xi) = 2\sum_{m=0}^{\infty} e^{-\eta_m^2(\theta_2(\mathrm{t})-\theta_2(\tau))}\sin\eta_m X \sin\eta_m\xi, \eta_m = \frac{2m+1}{2}\pi,$$
$$\mathcal{H}_2^{(2)}(t, \tau, \mathrm{X}, \xi) = 1 + 2\sum_{m=1}^{\infty} e^{-\eta_m^2(\theta_2(\mathrm{t})-\theta_2(\tau))}\cos\eta_m X \cdot \cos\eta_m\xi, \eta_m = m\pi,$$

where $\theta_2(t) = \int_0^t b(s)ds$.

The notation $\mathcal{H}_{4\tau\xi}^{(2)}, \mathcal{H}_{4\xi\xi}^{(2)}$ means partial derivatives of the influence function $\mathcal{H}_4^{(2)}$ relative to the definite variables τ and ξ, respectively.

Based on formula (20), we calculate

$$v_X(t, \mathrm{X}, \mathrm{Z}) = -\int_0^t \mathcal{H}_{4\xi X}^{(2)}(t, \tau, \mathrm{X}, 1)b(\tau)u(\tau, \mathrm{Z})d\tau \tag{21}$$

Integrating parts (21), taking into account the relations.

$$\mathcal{H}_{4X}^{(2)}(t, \tau, \mathrm{X}, \xi) = -\mathcal{H}_{3\xi}^{(2)}(t, \tau, X, \xi), \mathcal{H}_{3\tau}^{(2)}(t, \tau, X, \xi) = -b(\tau)\mathcal{H}_{3\xi\xi}^{(2)}(t, \tau, X, \xi),$$

and the initial condition $u_{|t=0} = 0$, we find

$$v_X(t, \mathrm{X}, \mathrm{Z}) = \int_0^t \mathcal{H}_3^{(2)}(t, \tau, \mathrm{X}, 1))u_\tau(\tau, \mathrm{Z})d\tau \tag{22}$$

We substitute the expression $v(t, \mathrm{X}, \mathrm{Z})$ (20) in the observation conditions (19):

$$u(t, h_i) - \int_0^1 XdX \int_0^t \mathcal{H}_{4\xi}^{(2)}(t, \tau, \mathrm{X}, 1)b(\tau)u(\tau, h_i)d\tau = \chi_i(\mathrm{t}), i = \overline{1, \mathrm{P}} \tag{23}$$

Integrating parts (23) and taking into account equality

$$\mathcal{H}_{4\xi}^{(2)}(t, \tau, X, 1) = -\mathcal{H}_{3X}^{(2)}(t, \tau, X, 1),$$

we obtain [16]

$$u(t,h_i) = \chi_i(\mathrm{t}) - \int_0^t \mathcal{H}_3^{(2)}(t,\tau,1,1)b(\tau)u(\tau,h_i)d\tau + \int_0^t\int_0^1 \mathcal{H}_3^{(2)}(t,\tau,\mathrm{X},1)b(\tau)u(\tau,h_i)dXd\tau, i = \overline{1,\mathrm{P}} \tag{24}$$

Let us first put $u(t,h_P) = \mu_{sP}(t) = C_s^{initial}(t)$, where $Z = h_P$ is the observation surface, approaching the point of entry into the work area Z = 1.

Then Eq. (24) for i = P will be

$$\int_0^t \mathcal{H}_3^{(2)}(t,\tau,1,1)b(\tau)\mu_{sP}(t)d\tau = \chi_{sP}(t) - \mu_{sP}(t) + \int_0^t\int_0^1 \mathcal{H}_3^{(2)}(t,\tau,\mathrm{x},1)b(\tau)\mu_{sP}(t)dxd\tau \tag{25}$$

Applying to Eq. (25) the formula $\int_\tau^t b(\sigma)\mathcal{H}_2^{(2)}(t,\sigma,0,0))\mathcal{H}_4^{(2)}(t,\sigma,0,0))d\sigma = 1$, obtained by Ivanchov [16], and taking into account.

$\mathcal{H}_3^{(2)}(t,\sigma;1,1) = \mathcal{H}_4^{(2)}(t,\sigma;0,0)$, we obtain

$$\int_0^t b(\tau)\mu_{sP}(\tau)d\tau = \int_0^t \mathcal{H}_2^{(2)}(t,\sigma,0,0)b(\sigma)(\chi_{sP}(\sigma) - \mu_{sP}(\sigma))d\sigma$$
$$+ \int_0^t \mathcal{H}_2^{(2)}(t,\sigma,0,0)b(\sigma)d\sigma \int_0^\sigma\int_0^1 \mathcal{H}_3^{(2)}(\sigma,\tau,\mathrm{x},1)b(\tau)\mu_{sP}(t)dXd\tau, t \in \left[0,t^{total}\right] \tag{26}$$

Differentiating Eq. (26) by t, after the transformations series, we obtain

$$\mu_{sP}(t) = \int_0^t \mathcal{H}_2^{(2)}(t,\sigma,0,0)(b(\sigma)\mu_{sP}(\sigma) + \chi_{sP}(\sigma) - \mu_{sP}(\sigma))d\sigma \tag{27}$$

After multiplying Eq. (27) on the expression $\mathcal{H}_4^{(2)}(t,\sigma,0,0))b(\sigma)$, the integration by τ and the differentiation by t

$$b(t)\mu_{sP}(t) + \chi_{sP}(\sigma) - \mu_{sP}(\sigma) = b(t)\int_0^t \mathcal{H}_2^{(2)}(t,\tau,0,0))\mu'_{sP}(\tau))d\tau.$$

So we obtain the expression for calculating the co-diffusion coefficient in the intra-crystallite space:

$$D_{\mathrm{intra}_{sP}}(t) \equiv R^2 b(t) = R^2 \frac{\chi'_{sP}(t) - \mu'_{sP}(t)}{\int_0^t \mathcal{H}_2^{(2)}(t,\tau,0,0))\mu'_{sP}(\tau)d\tau - \mu_{sP}(t)}, t \in \left[0,t^{total}\right] \tag{28}$$

Using calculated $D_{\mathrm{intra}_{sP}}(t)$ with the formula (28) on the observation limit h_P, we define the gradient method $D_{\mathrm{inter}_{sP}}(t)$ in the same way. With $D_{\mathrm{intra}_{sP}}(t)$ and $D_{\mathrm{inter}_{sP}}(t)$ in h_P, we calculate $C_{sk}(t,h_P)$, substituting it in $\mu_{sP-1}(t) = C_{sk}(t,h_P)$ for the next coefficient $D_{\text{inter}_{si}}(t), i = \overline{P-1,1}$ calculations. All subsequent coefficients $D_{\text{intra}_{si}}(t)$ will be calculated by the formula

$$D_{\text{intra}_{si}}(t) \equiv R^2 b_{si}(t) = R^2 \frac{\chi'_{si}(t) - \mu'_{si}(t)}{\int_0^t \mathcal{H}_2^{(2)}(t,\tau,0,0))\mu'_{si}(\tau)d\tau - \mu_{si}(t)}, i = \overline{P-1,1} \tag{29}$$

with parallel computing $D_{\text{inter}_{si}}(t), i = \overline{P-1,1}$.

5. Numerical simulation and analysis: co-diffusion coefficients: concentration profiles in inter- and intra-crystallite spaces

The variation against time of the benzene and hexane intra-crystallite diffusion coefficients, $D_{\text{intra}_{1,k}}$ and $D_{\text{intra}_{2,k}}$, respectively, is presented in **Figure 3** for the five coordinate positions: 6, 8, 10, 12, and 14 mm, defined now from the top of the bed. The curves for benzene $D_{\text{intra}_{1,k}}$ are pseudo exponentials. $D_{\text{intra}_{1,k}}$ decreases from 9.0

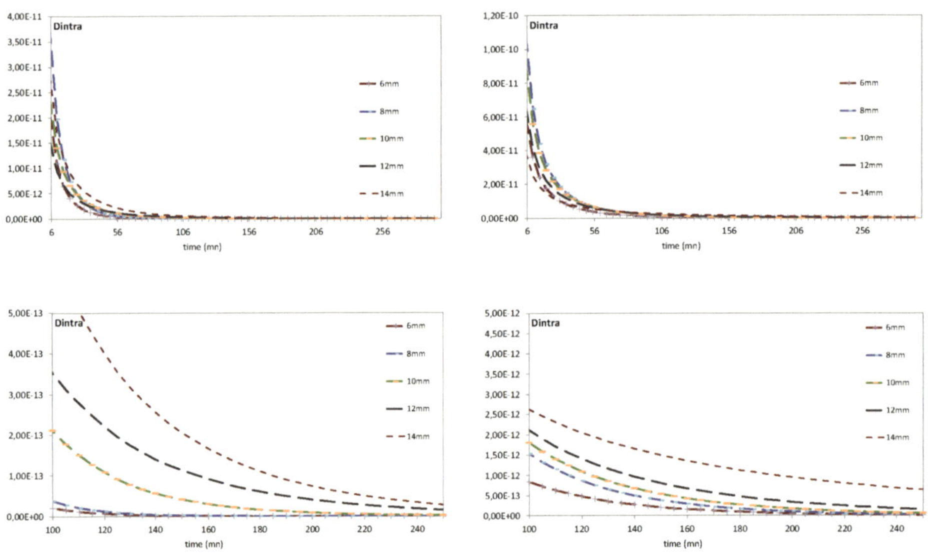

Figure 3.
Variation of intra-crystallite diffusion coefficients (arbitrary units) for benzene $D_{intra_{1,k}}$ (left) and hexane $D_{intra_{2,k}}$ (right) against time, at different positions in the bed. (Top) time range 6–240 mn, (bottom) time range 100–240 mn.

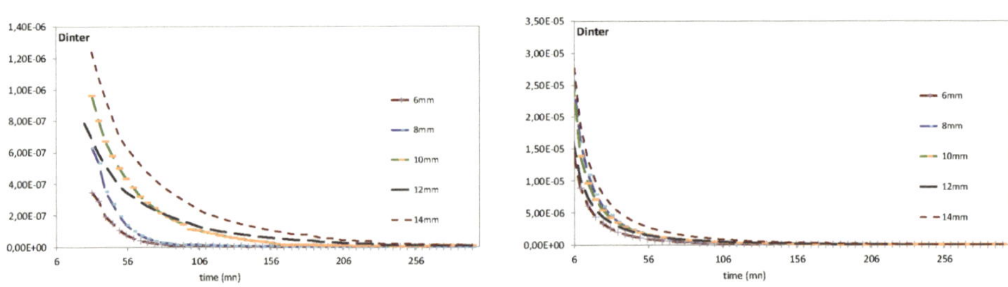

Figure 4.
Variation of inter-crystallite diffusion coefficients (a.u.) for benzene (left) and hexane (right) against time at different positions in the bed.

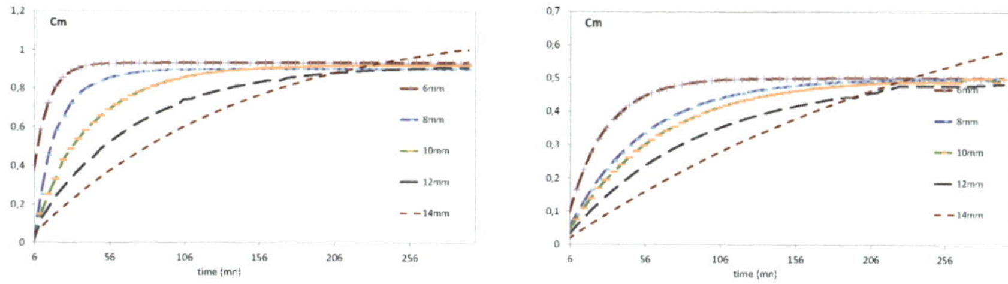

Figure 5.
Variation of the inter-crystallite concentration (a.u.) calculated for benzene (left) and hexane (right) against time and at different positions in the bed.

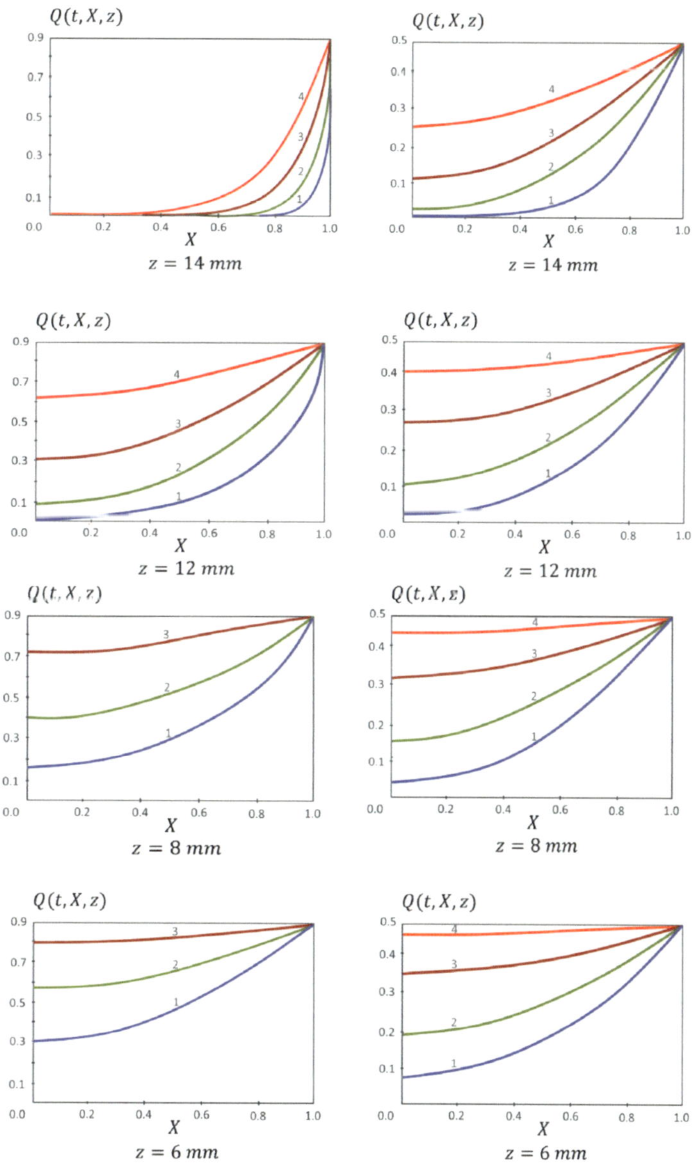

Figure 6.
Distribution of the benzene (left) and hexane (right) concentrations in the intra-crystallite space from the surface (abscissa 1) to the center (abscissa 0) of the crystallites, at different times: (1) dark blue, t = 25 min; (2) green, t = 50 min; (3) brown, t = 100 min; and (4) red, t = 200 min.

E−13 to about 1.0 E−14 a.u. (equilibrium) depending on the position of the crystallite and the time, as well as on the amount of adsorbed gas. The shapes of the variations of $D_{intra_{2,k}}$ for hexane are roughly the same, but the diffusion coefficients are higher, from about 9.0 E−12 to 3.0 E−13 a.u.

Figure 4 presents the variation against time of the benzene and hexane diffusion coefficients in inter-crystallite space, $D_{inter_{1,k}}$ and $D_{inter_{2,k}}$, for the same positions. These coefficients decrease with time from 2.0 E−6 to 1.0 E−7 a.u. (equilibrium) for benzene and from 3 E−5 to 1.0 E−6 a.u. for hexane, depending on the bed position, and increase adsorbed concentrations.

Figure 5 shows the variation against time of the calculated concentrations C for benzene and hexane in the inter-crystallite space. As can be seen, these concentrations approach the equilibrium values for a diffusion time around 250 min. But the variations of the concentrations with time are rather different for the two gases.

Figure 6 shows the variation of the concentrations $Q(t,X,z)$ of adsorbed benzene (left) and hexane (right) in the micropores of the intra-crystallite space from the surface (abscissa-1) to the center (abscissa-0) of the crystallites located between 6 and 14 mm from the top of the bed and after 25–200 min. of diffusion (a, b, c, and d, respectively). The gradients increase, and the mean concentrations decrease with the increasing distance of the particles from the arrival of the gases. The particles at 6 and 8 mm are saturated with benzene after 100 min, but not yet with hexane.

6. Conclusion

The main result of this work is the possibility, from a single experiment, of simultaneously distributing several co-diffusing gases in a porous solid and of using the methods of mathematical modeling to analyze for each of them the distribution of their concentrations in the intra- and inter-crystallite spaces.

Using the experimental NMR data and proposed co-adsorption models, the identification procedures for calculating the co-diffusion coefficients for two or more components in intra- and inter-crystallite spaces are developed. These procedures use the iterative gradual identification methods on minimizing of the Lagrange error function and rapid analytic methods based on the influence function. The co-diffusion coefficients were obtained as a function of time for different positions along the catalyst bed. In particular, those in the intra-crystallite space were computed by the analytical method which allowed a calculation with a relatively high degree of discretization over time and to reduce practically twice the volume of iterative calculations. Using these results, the concentrations of co-diffusing benzene and hexane in the inter- and intra-crystallite spaces were calculated for each time and each position in the bed.

Nomenclature

$k = \overline{1, N+1}$	layer number, subscript k will be added to all the following symbols to specify that they are characteristic of the kth layer
c	adsorbate concentration in macropores
c_∞	equilibrium adsorbate concentration in macropores
$C = c/c_\infty$	dimensionless adsorbate concentration in macropores

D_{inter}	diffusion coefficient in macropores, m^2/s
D_{intra}	diffusion coefficient in micropores, m^2/s
$\tilde{K}$	adsorption equilibrium constant
l	bed length, m
$\Delta l = l_k - l_{k-1}; k = \overline{1, N+1}$	layer thickness (all layers have the same thickness)
L	dimensionless bed length (L = 1)
q	adsorbate concentration in micropores
q_∞	equilibrium adsorbate concentration in micropores
$Q = q/q_\infty$	dimensionless concentration of adsorbate in micropores
T	temperature of gas phase flow, K, and time total, s
M	mass total
u	velocity of gas phase flow, m/s^2
Λ	coefficient of thermal diffusion along the columns
h_g	gas heat capacity, kJ/(kg K)
μ	molecular mass of adsorbate, kg/mol
H	total heat capacity of the adsorbent and gas, kJ/(kg K)
α_h	heat transfer coefficient
R_{column}	column radius, m
R_g	gas constant, kJ mol/(m^3 K)
ΔH_i	activation energy ($\Delta H_i = \Delta \overline{H}_i/\mu$), kJ/mol
$\Delta \overline{H}_i$	adsorption heat, kJ/kg
k_{0i}	empirical equilibrium coefficient for the *i* adsorbate, depending on the adsorbent properties and the diffusing adsorbate component (k_{0i} equal to the ratio of the desorption and adsorption rate constants)
x	distance from crystallite center, m
R	mean crystallite radius, m (we assume that the crystallites are spherical)
$X = x/R$	dimensionless distance from crystallite center
z	distance from the bottom of the bed for mathematical simulation, m
$Z = z/l$	dimensionless distance from the bottom of the bed
t	time
τ, ξ	variables of integration
t^{total}	total duration of co-adsorption, mn
L_k	dimensionless position of the *k*th layer
h_k	$(L_k - L_{k-1})/2$
ε_{inter}	inter-crystallite bed porosity
e_{inter}	value utilized in Eq. (9)
n	iteration number of identification
m	number of adsorbed components
P	number of NMR observation surfaces
s	index of adsorbate component
i	index of NMR observation surface
initial	index of initial value (concentrations, temperature)
macro	index of extended Lagrange functional component for inter-crystallite space
micro	index of extended Lagrange functional component for intra-crystallite space

Appendix

A. Iterative gradient method of the identification of co-diffusion coefficients

The methodology for solving the direct boundary problem (9)–(15), which describes the diffusion process in a heterogeneous nanoporous bed, is developed in [9, 12, 15]. According to [12] the procedure for determining the diffusion coefficients (16) requires a special technique for calculating the gradients $\nabla J^n_{D_{\text{intra}_{s_k}}}(t)$ and $\nabla J^n_{D_{\text{inter}_{s_k}}}(t)$ of the residual functional (17). This leads to the problem of optimizing the extended Lagrange function [12, 15]:

$$\Phi\left(D_{\text{inter}_{sk}}, D_{\text{intra}_{sk}}\right) = J_s + I_{s_{macro}} + I_{s_{micro}}, \tag{A.1}$$

where $I_{s_{macro}}, I_{s_{micro}}$ are the components given by Eqs. (A.2) and (A.3), corresponding to the macro- and microporosity, respectively

$$I_{s_{macro}} = \int_0^T \int_{L_{k-1}}^{L_k} \phi_{s_k}(t,Z)\left(\frac{\partial C_{s_k}}{\partial t} - \frac{D_{\text{inter}_{sk}}}{l^2}\frac{\partial^2 C_{s_k}}{\partial Z^2} + e_{\text{inter}_k} K_{s_k} \frac{D_{\text{intra}_{s_k}}}{R^2}\left(\frac{\partial Q(t,X,Z)_{s_k}}{\partial X}\right)_{X=1}\right) dZdt \tag{A.2}$$

$$I_{s_{micro}} = \int_0^T \int_0^1 \int_{L_{k-1}}^{L_k} \psi_{s_k}(t,X,Z)\left(\frac{\partial Q_{s_k}(t,X,Z)}{\partial t} - \frac{D_{\text{intra}_{sk}}}{R^2}\left(\frac{\partial^2 Q_{s_k}}{\partial X^2} + \frac{2}{X}\frac{\partial Q_{s_k}}{\partial X}\right)\right) XdXdZdt \tag{A.3}$$

J_s is the residual functional (17), $\phi_{s_k}, \psi_{s_k}, {}_{s,=\overline{1,2}}$—unknown factors of Lagrange, to be determined from the stationary condition of the functional $\Phi\left(D_{\text{inter}_{sk}}, D_{\text{intra}_{sk}}\right)$ [9, 15]:

$$\Delta\Phi\left(D_{\text{inter}_{sk}}, D_{\text{intra}_{sk}}\right) \equiv \Delta J_s + \Delta I_{s_{macro}} + \Delta I_{s_{micro}} = 0 \tag{A.4}$$

The calculation of the components in Eq. (A.4) is carried out by assuming that the values $D_{\text{inter}_{sk}}, D_{\text{intra}_{sk}}$ are incremented by $\Delta D_{\text{inter}_{sk}}, \Delta D_{\text{intra}_{sk}}$. As a result, concentration $C_{s_k}(t,Z)$ changes by increment $\Delta C_{sk}(t,Z)$ and concentration $Q_{s_k}(t,X,Z)$ by increment $\Delta Q_{s_k}(t,X,Z), s = \overline{1,2}$.

Conjugate problem. The calculation of the increments ΔJ_s, $\Delta J_{s_{macro}}$, and $\Delta J_{s_{micro}}$ in Eq. (A.4) (using integration by parts and the initial and boundary conditions of the direct problem (9)–(15)) leads to solving the additional conjugate problem to determine the Lagrange factors ϕ_{s_k} and ψ_{s_k} of the functional (A.1) [15]:

$$\frac{\partial \phi_{s_k}(t,Z)}{\partial t} + \frac{D_{\text{inter}_{sk}}}{l^2}\frac{\partial^2 \phi_{s_k}}{\partial Z^2} + e_{\text{inter}_k} K_{s_k} \frac{D_{\text{intra}_{s_k}}}{R^2}\frac{\partial \psi_{s_k}(t,X,Z)}{\partial X}\bigg|_{X=1} = E^n_{s_k}(t)\delta(Z-h_k) \tag{A.5}$$

where $E^n_{s_k}(t) = C_{s_k}\left(D_{\text{intra}_{sk}n}, D_{\text{inter}_{sk}n}; t, h_k\right) + \overline{Q}_{s_k}\left(D_{\text{intra}_{sk}n}, D_{\text{inter}_{sk}n}; t, h_k\right) - M_{sk}(t)$, $\delta(Z-h_k)$ (function of Dirac) [15].

$$\frac{\partial \psi_{s_k}(t,X,Z)}{\partial t} + \frac{D_{\text{intra}_{sk}}}{R^2}\left(\frac{\partial^2 \psi_{s_k}}{\partial X^2} + \frac{2}{X}\frac{\partial \psi_{s_k}}{\partial X}\right) = E^n_{s_k}(t)\delta(Z-h_k) \tag{A.6}$$

$$\phi_{s_k}(t,Z)_{|t=t^{total}} = 0; \psi_{s_k}(t,X,Z)_{|t=t^{total}} = 0 \ \left(\text{conditions at } t = t^{total}\right); \tag{A.7}$$

$$\frac{\partial}{\partial X}\psi_{s_k}(t,X,Z)_{|X=0} = 0; \psi_{s_k}(t,X,Z)_{|X=1} = \varphi_{s_k}(t,Z) \tag{A.8}$$

$$\phi_{s_k}(t,Z=L_k) = 0, \phi_{s_{k-1}}(t,Z=L_{k-1}) = 0; s=\overline{1,2}, k=\overline{N,2}, \tag{A.9}$$

$$\phi_{s_1}(t,L_1) = 0, \ \frac{\partial \phi_{s_1}}{\partial Z}(t,Z=0) = 0 \tag{A.10}$$

We have obtained the solutions ϕ_{sk} and ψ_{sk} of problem (A.5)–(A.10) using Heaviside operational method in [15].

Substituting in the direct problem (9)–(15) $D_{\text{inter}_{\text{sk}}}$, $D_{\text{intra}_{\text{sk}}}$, $C_{s_k}(t,Z)$, and $Q_{s_k}(t,X,Z)$ by the corresponding values with increments $D_{\text{inter}_{\text{sk}}} + \Delta D_{\text{inter}_{\text{sk}}}$, $D_{\text{intra}_{\text{sk}}} + \Delta D_{\text{intra}_{\text{sk}}}$, $C_{s_k}(t,Z) + \Delta C_{s_k}(t,z)$, and $Q_{s_k}(t,X,Z) + \Delta Q_{s_k}(t,X,Z)$, subtracting the first equations from the transformed ones and neglecting second-order terms of smallness, we obtain the basic equations of the problem (9)–(15) in terms of increments $\Delta C_{s_k}(t,Z)$ and $\Delta Q_{s_k}(t,X,Z), s=\overline{1,2}$ in the operator form

$$\mathcal{L} w_{s_k}(t,X,Z) = \mathrm{X}_{s_k}, w_{s_k} \in (0,1) \cup \Omega_{kt}, k = \overline{1, N+1} \tag{A.11}$$

Similarly, we write the system of the basic equations of conjugate boundary problem (A.5)–(A.10) in the operator:

$$\mathcal{L}^* \Psi_{s_k}(t,X,Z) = E_{s_k}(t)\delta(Z-h_k), \Psi_{s_k} \in (0,1) \cup \Omega_{kt}, k = \overline{1, N+1} \tag{A.12}$$

where $\mathcal{L} = \begin{bmatrix} \frac{\partial}{\partial t} - \frac{\partial}{\partial Z}\left(D_{\text{inter}_{\text{sk}}}\frac{\partial}{\partial Z}\right) & e_{\text{inter}_k}\frac{D_{\text{intra}_{\text{sk}}}}{R}\frac{\partial}{\partial X}|_{X=1} \\ 0 & \frac{\partial}{\partial t} - \frac{D_{\text{intra}_{\text{sk}}}}{R^2}\left(\frac{\partial^2}{\partial X^2} + \frac{2}{X}\frac{\partial}{\partial X}\right) \end{bmatrix}$,

$$\mathcal{L}^* = \begin{bmatrix} \frac{\partial}{\partial t} + \frac{\partial}{\partial Z}\left(D_{\text{inter}_{\text{sk}}}\frac{\partial}{\partial Z}\right) & e_{\text{inter}_k}\frac{D_{\text{intra}_{\text{sk}}}}{R^2}\frac{\partial}{\partial X}|_{X=1} \\ 0 & \frac{\partial}{\partial t} + \frac{D_{\text{intra}_{\text{sk}}}}{R^2}\left(\frac{\partial^2}{\partial X^2} + \frac{2}{X}\frac{\partial}{\partial X}\right) \end{bmatrix},$$

$$w_{s_k}(t,X,Z) = \begin{bmatrix} \Delta C_{s_k}(t,Z) \\ \Delta Q_{s_k}(t,X,Z) \end{bmatrix}, \Psi_{s_k}(t,X,Z) = \begin{bmatrix} \phi_{s_k}(t,Z) \\ \psi_{s_k}(t,X,Z) \end{bmatrix}.$$

$$\mathrm{X}_{s_k}(t,X,Z) = \begin{bmatrix} \frac{\partial}{\partial Z}\left(\Delta D_{\text{inter}_{\text{sk}}}\frac{\partial}{\partial Z}C_{s_k}\right) - e_{\text{inter}_k}\frac{\Delta D_{\text{intra}_{\text{sk}}}}{R^2}\frac{\partial}{\partial X}Q_{s_k}(t,X,Z)_{X=1} \\ \frac{\Delta D_{\text{intra}_{\text{sk}}}}{R^2}\left(\frac{\partial^2}{\partial X^2} + \frac{2}{X}\frac{\partial}{\partial X}\right)Q_{s_k}(t,X,Z) \end{bmatrix} \tag{A.13}$$

where $\mathcal{L}^*$ is the conjugate Lagrange operator of operator $\mathcal{L}$.

The calculated increment of the residual functional (17), neglecting second-order terms, has the form

$$\begin{aligned} \Delta J_s\left(D_{\text{intra}_{\text{sk}}}, D_{\text{inter}_{\text{sk}}}\right) &= \int_0^T \int_{L_{k-1}}^{L_k} \mathcal{L}^{-1}\mathrm{X}_{s_{k,1}}(t,Z) \cdot E_{s_k}(t)\delta(Z-h_k)dZdt \\ &+ \int_0^T \int_{L_{k-1}}^{L_k} \int_0^1 \mathcal{L}^{-1}\mathrm{X}_{s_{k2}}(t,X,Z) \cdot E_{s_k}(t)\delta(Z-h_k)XdXdZdt \end{aligned} \tag{A.14}$$

where $w_{s_k} = \mathcal{L}^{-1}X_{s_k}$ and $\mathcal{L}^{-1}$ is the inverse operator of operator $\mathcal{L}$.
Defining the scalar product

$$\left(\mathcal{L}w_{s_k}(t,X,Z), \Psi_{s_k}(t,X,Z)\right) = \begin{bmatrix} \int\int\limits_{\Omega_{kT}} \mathcal{L}\Delta C_{s_k}(t,Z)\phi_{s_k}(t,Z)dZdt \\ \iiint\limits_{(0,R)\cup\Omega_{kT}} \mathcal{L}\Delta Q_{s_k}(t,X,Z)\psi_{s_k}(t,X,Z)XdXdZdt \end{bmatrix} \tag{A.15}$$

and taking into account (A.19) Lagrange's identity [12, 15]

$$\left(\mathcal{L}w_{s_k}(t,X,Z), \Psi_{s_k}(t,X,Z)\right) = \left(w_{s_k}(t,X,Z), \mathcal{L}^*\Psi_{s_k}(t,X,Z)\right) \tag{A.16}$$

and the equality $\mathcal{L}^{-1*}\left[E_{s_k}(t)\delta(Z-h_k)\right] = \Psi_{s_k}$, we obtain the increment of the residual functional expressed by the solution of conjugate problem (A.6)–(A.10) and the vector of the right-hand parts of Eq. (A.13):

$$\Delta J_s\left(D_{\text{inter}_{sk}}, D_{\text{intra}_{sk}}\right) = \left(\Psi_{s_k}(t,X,Z), X_{s_k}(t,X,Z)\right) \tag{A.17}$$

where $\phi_{s_k}(t,Z)$ and $\psi_{s_k}(t,X,Z)$ belong to $\overline{\Omega}_{kt}$ and $[0,1]\cup\overline{\Omega}_{kt}$, respectively, $\mathcal{L}^{-1*}$ is the conjugate operator to inverse operator $\mathcal{L}^{-1}$, and Ψ_{s_k} is the solution of conjugate problem (A.5)–(A.10).

Reporting in Eq. (A.17) the components $X_{s_k}(t,X,Z)$ taking into account Eq. (A.18), we obtain the formula which establishes the relationship between the direct problem (9)–(15) and the conjugate problem (A.6)–(A.10) which makes it possible to obtain the analytical expressions of components of the residual functional gradient:

$$\Delta J_s\left(D_{\text{intra}_{sk}}, D_{\text{inter}_{sk}}\right) = \begin{matrix} \left(\phi_{s_k}(t,Z), \frac{\partial}{\partial Z}\left(\Delta D_{\text{inter}_{sk}}\frac{\partial}{\partial Z}C_{s_k}\right) - e_{\text{inter}_k}\frac{\Delta D_{\text{intra}_{sk}}}{R^2}\frac{\partial}{\partial X}Q_{s_k}(t,X,Z)_{X=1}\right) \\ +\left(\psi_{s_k}(t,X,Z), \frac{\Delta D_{\text{intra}_{sk}}}{R^2}\left(\frac{\partial^2}{\partial X^2}+\frac{2}{X}\frac{\partial}{\partial X}\right)Q_{s_k}(t,X,Z)\right) \end{matrix} \tag{A.18}$$

Differentiating expression (A.18), by $\Delta D_{\text{intra}_{sk}}$ and $\Delta D_{\text{inter}_{sk}}$, respectively, and calculating the scalar products according to Eq. (A.15), we obtain the required analytical expressions for the gradient of the residual functional in the intra- and inter-crystallite spaces, respectively:

$$\nabla J_{D_{\text{intra}_{sk}}}(t) = -\frac{e_{\text{inter}_k}}{R^2}\int\limits_{L_{k-1}}^{L_k}\frac{\partial}{\partial X}Q_{s_k}(t,1,Z)\phi_{s_k}(t,Z)dZ + \frac{1}{R^2}\int\limits_{L_{k-1}}^{L_k}\int\limits_0^1\left(\frac{\partial^2}{\partial X^2}+\frac{2}{X}\frac{\partial}{\partial X}\right)Q_{s_k}(t,X,Z)\psi_{s_k}(t,X,Z)XdXdZ \tag{A.19}$$

$$\nabla J_{D_{\text{inter}_{sk}}}(t) = \int\limits_{L_{k-1}}^{L_k}\frac{\partial^2 C_{s_k}(t,Z)}{\partial Z^2}\phi_{s_k}(t,Z)dZ \tag{A.20}$$

The formulas of gradients $\nabla J^n_{D_{\text{intra}_{s_k}}}(t)$ and $\nabla J^n_{D_{\text{inter}_{s_k}}}(t)$ include analytical expressions of the solutions of the direct problem (9)–(14) and inverse problem

(A.5)–(A.10). They provide high performance of computing process, avoiding a large number of inner loop iterations by using exact analytical methods [2, 15].

B. The linearization schema of the nonlinear co-adsorption model: system of linearized problems and construction of solutions

The linearization schema of nonlinear co-adsorption (1)–(8) is shown in order to demonstrate the simplicity of implementation for the case of two diffusing components ($m = 2$) and isothermal adsorption. The simplified model (1)–(8) for the case of $m = 2$ is converted into the form

$$\frac{\partial C_s(t,Z)}{\partial t} = \frac{D_{\text{inter}_s}}{l^2}\frac{\partial^2 C_s}{\partial Z^2} - e_{\text{inter}}\tilde{K}_s\frac{D_{\text{intra}_s}}{R^2}\left(\frac{\partial Q_s}{\partial X}\right)_{X=1} \tag{A.21}$$

$$\frac{\partial Q_s(t,X,Z)}{\partial t} = \frac{D_{\text{intra}_s}}{R^2}\left(\frac{\partial^2 Q_s}{\partial X^2} + \frac{2}{X}\frac{\partial Q_s}{\partial X}\right), s = \overline{1,2} \tag{A.22}$$

with initial conditions

$$C_s(t=0,Z) = 0; Q_s(t=0,X,Z) = 0; X \in (0,1), s = \overline{1,2} \tag{A.23}$$

boundary conditions for coordinate X of the crystallite

$$\frac{\partial}{\partial X}Q_s(t,X=0,Z) = 0 \tag{A.24}$$

$$Q_1(t,X=1,Z) = \frac{K_1C_2(t,Z)}{1+K_1C_1(t,Z)+K_2C_2(t,Z)} \quad \text{(Langmuir equilibrium)},$$
$$Q_2(t,X=1,Z) = \frac{K_2C_2(t,Z)}{1+K_1C_1(t,Z)+K_2C_2(t,Z)} \tag{A.25}$$

boundary and interface conditions for coordinate Z

$$C_s(t,1) = 1, \frac{\partial C_s}{\partial Z}(t,Z=0) = 0, t \in \left(0, t^{total}\right) \tag{A.26}$$

$K_1 = \frac{\bar{\theta}_1}{p_1\left(1-\bar{\theta}_1-\bar{\theta}_2\right)}, K_2 = \frac{\bar{\theta}_2}{p_2\left(1-\bar{\theta}_1-\bar{\theta}_2\right)}$, where p_1, p_2 are the co-adsorption equilibrium constants and partial pressure of the gas phase for 1-th and 2-th component and $\bar{\theta}_1, \bar{\theta}_2$, are the intra-crystallite spaces occupied by the corresponding adsorbed molecules. The expression $\varphi_s(C_1, C_2) = \frac{C_s(t,Z)}{1+K_1C_1(t,Z)+K_2C_2(t,Z)}$ is represented by the series of Tailor [5]:

$$\varphi_s(C_1,C_2) = \varphi_s(0,0) + \left(\left.\frac{\partial\varphi_s}{\partial C_1}\right|_{(0,0)} C_1 + \left.\frac{\partial\varphi_s}{\partial C_2}\right|_{(0,0)} C_2\right) + \frac{1}{2!}\left(\left.\frac{\partial^2\varphi_s}{\partial C_1^2}\right|_{(0,0)} C_1^2 + 2\left.\frac{\partial^2\varphi_s}{\partial C_1 \partial C_2}\right|_{(0,0)} C_1C_1 + \left.\frac{\partial^2\varphi_s}{\partial C_2^2}\right|_{(0,0)} C_2^2\right) + \ldots \tag{A.27}$$

As a result of transformations, limiting to the series not higher than the second order, we obtain

$$\frac{K_1C_1(t,Z)}{1+K_1C_1(t,Z)+K_2C_2(t,Z)} = K_1C_1(t,Z) - \left(K_1^2C_1^2(t,Z) + \frac{1}{2}K_1K_2C_1(t,Z)C_2(t,Z)\right),$$
$$\frac{K_2C_2(t,Z)}{1+K_1C_1(t,Z)+K_2C_2(t,Z)} = K_2C_2(t,Z) - \left(\frac{1}{2}K_1K_2C_1^2(t,Z)C_2^1(t,Z) + K_2^2C_2^2(t,Z)\right) \tag{A.28}$$

Substituting the expanded expression (A.28) in Eq. (A.25) of nonlinear systems (A.20)–(A.26), we obtain

$$Q_1(t,X,Z)_{X=1} = K_1C_1(t,Z) - \varepsilon\left(C_1^2(t,Z) + \frac{1}{2}\frac{K_2}{K_1}C_1(t,Z)C_2(t,Z)\right),$$
$$Q_2(t,X,Z)_{X=1} = K_2C_2(t,Z) - \varepsilon\left(\frac{1}{2}\frac{K_2}{K_1}C_1(t,Z)C_2(t,Z) + \left(\frac{K_2}{K_1}\right)^2C_2^2(t,Z)\right) \tag{A.29}$$

where $\varepsilon = K_1^2 < < 1$ is the small parameter.

Taking into account the approximate equations of the kinetics of co-adsorption (A.29) containing the small parameter ε, we search for the solution of the problem (A.21)–(A.26) by using asymptotic series with a parameter ε in the form [6, 7]

$$C_s(t,Z) = C_{s_0}(t,Z) + \varepsilon C_{s_1}(t,Z) + \varepsilon^2 C_{s_2}(t,Z) + \ldots, \tag{A.30}$$

$$Q_s(t,X,Z) = Q_{s_0}(t,X,Z) + \varepsilon Q_{s_1}(t,X,Z) + \varepsilon^2 Q_{s_2}(t,X,Z) + \ldots, s = \overline{1,2} \tag{A.31}$$

As the result of substituting the asymptotic series (A.30)–(A.31) into the equations of the nonlinear boundary problem (A.21)–(A.26) considering Eq. (A.28), the problem (A.21)–(A.26) will be parallelized into two types of linearized boundary problems [6]:

The problem $A_{s_0}, s = \overline{1,2}$ (*zero approximation with initial and boundary conditions of the initial problem*): to find a solution in the domain $D = \{(t,X,Z) : t>0, X\in(0,1), Z\in(0,1)\}$ of a system of partial differential equations

$$\frac{\partial C_{s0}(t,Z)}{\partial t} = \frac{D_{\text{inter}_s}}{l^2}\frac{\partial^2 C_{s0}}{\partial Z^2} - e_{\text{inter}}\tilde{K}_s\frac{D_{\text{intra}_s}}{R^2}\left(\frac{\partial Q_{s0}}{\partial X}\right)_{X=1} \tag{A.32}$$

$$\frac{\partial Q_{s_0}(t,X,Z)}{\partial t} = \frac{D_{\text{intra}_s}}{R^2}\left(\frac{\partial^2 Q_{s0}}{\partial X^2} + \frac{2}{X}\frac{\partial Q_{s0}}{\partial X}\right) \tag{A.33}$$

with initial conditions.

$$C_{s_0}(t=0,Z) = 0; Q_{s_0}(t=0,X,Z) = 0; X\in(0,1), s = \overline{1,2} \tag{A.34}$$

boundary conditions for coordinate X of the crystallite

$$\frac{\partial}{\partial X}Q_{s_0}(t,X=0,Z) = 0 \tag{A.35}$$

$$Q_{s_0}(t,X=1,Z) = K_sC_{s_0}(t,Z), s = \overline{1,2} \tag{A.36}$$

boundary and interface conditions for coordinate Z.

$$C_{s_0}(t,1) = 1, \frac{\partial C_{s_0}}{\partial Z}(t,Z=0) = 0, t \in (0,T) \tag{A.37}$$

The problem $A_n; n = \overline{1,\infty}$ (nth *approximation with zero initial and boundary conditions*): to construct in the domain D a solution of a system of equations

$$\frac{\partial C_{s_n}(t,Z)}{\partial t} = \frac{D_{\text{inter}_s}}{l^2}\frac{\partial^2 C_{s_n}}{\partial Z^2} - e_{\text{inter}}\tilde{K}_s \frac{D_{\text{intra}_s}}{R^2}\left(\frac{\partial Q_{s_n}}{\partial X}\right)_{X=1} \tag{A.38}$$

$$\frac{\partial Q_{s_n}(t,X,Z)}{\partial t} = \frac{D_{\text{intra}_s}}{R^2}\left(\frac{\partial^2 Q_{s_n}}{\partial X^2} + \frac{2}{X}\frac{\partial Q_{s_n}}{\partial X}\right) \tag{A.39}$$

with initial conditions.

$$C_{s_n}(t=0,Z) = 0; Q_{s_n}(t=0,X,Z) = 0; s = \overline{1,2} \tag{A.40}$$

boundary conditions for coordinate X of the crystallite.

$$\frac{\partial}{\partial X}Q_{s_n}(t,X=0,Z) = 0 \tag{A.41}$$

$$Q_{1_n}(t,X,Z)_{X=1} = K_1 C_{1_n}(t,Z) - \sum_{\nu=0}^{n-1} C_{1_\nu}(t,\mathrm{Z})\left(C_{1,n-1-\nu}(t,\mathrm{Z}) + \frac{1}{2}\frac{K_2}{K_1}C_{2,n-1-\nu}(t,\mathrm{Z})\right),$$

$$Q_{2_n}(t,X,Z)_{X=1} = K_2 C_{2_n}(t,Z) - \sum_{\nu=0}^{n-1} C_{2_s}(t,\mathrm{Z})\left(\frac{1}{2}\frac{K_2}{K_1}C_{1,n-1-\nu}(t,\mathrm{Z}) + \left(\frac{K_2}{K_1}\right)^2 C_{2,n-1-\nu}(t,\mathrm{Z})\right) \tag{A.42}$$

boundary and interface conditions for coordinate Z.

$$C_{s_n}(t,1) = 0, \frac{\partial C_{s_0}}{\partial Z}(t,Z=0) = 0, t \in \left(0, t^{total}\right) \tag{A.43}$$

The problems $A_{s_0}, s = \overline{1,2}$ are linear with respect to zero approximation C_{s_0}, Q_{s_0}; the problems A_{s_n}; n $= \overline{1,\infty}$ are linear with respect to the n-th approximation C_{s_n}, Q_{s_n} and nonlinear with respect to all previous *n-1* approximations $C_{s_0}, \ldots, C_{s_{n-1}}$.

As demonstrated for the two-component adsorption model (A.21)–(A.26), our proposed methodology can easily be developed and applied to the co-adsorption of any number of gases.

Author details

Mykhaylo Petryk[1*], Mykola Ivanchov[2], Sebastian Leclerc[3], Daniel Canet[3] and Jacques Fraissard[4]

1 Ternopil Ivan Puluj National Technical University, Ternopil, Ukraine

2 Ivan Franco National University of Lviv, Lviv, Ukraine

3 University of Lorraine, Vandoeuvre-les-Nancy, France

4 Faculty of Science and Engineering, ESPCI, Sorbonne University, Paris, France

*Address all correspondence to: mykhaylo_petryk@tu.edu.te.ua

References

[1] Leclerc S, Trausch G, Cordier B, Grandclaude D, Retournard A, Fraissard J, et al. Chemical shift imaging (CSI) by precise object displacement. Magnetic Resonance in Chemistry. 2006;**44**:311-317

[2] Petryk M, Leclerc S, Canet D, Fraissard J. Mathematical modeling and visualization of gas transport in a zeolite bed using a slice selection procedure. Diffusion Fundamentals. 2007;**4**:11.1. Available from: http://www.diffusion-fundamentals.org

[3] Petryk M, Leclerc S, Canet D, Fraissard J. Modeling of gas transport in a microporous solid using a sclice selection procedure: Application to the diffusion of benzene in ZSM5. Catalysis Today. 2008;**139**:234-240

[4] Leclerc S, Petryk M, Canet D, Fraissard J. Competitive diffusion of gases in a zeolite using proton NMR and a slice selection procedure. Catalysis Today. 2012;**187**:104-107

[5] Petryk MR, Khimich OM, Boyko IV, Mykhalyk DM, Petryk MM, Kovbashyn VI. Mathematical Modeling of Heat Transfer and Adsorption of Hydrocarbons in Nanoporous Media of Exhaust Gas Neutralization Systems. Kyiv: National Academy of Sciences of Ukraine; 2018. p. 280

[6] Petryk M, Khimitch A, Petryk MM. Simulation of adsorption and desorption of hydrocarbons in nanoporous catalysts of neutralization systems of exhaust gases using nonlinear langmuir isotherm. Journal of Automation and Information Sciences, Begell House USA. 2018;**50**(10):18-33

[7] Petryk M, Khimitch A, Petryk MM, Fraissard J. Experimental and computer simulation studies of dehydration on microporous adsorbent of natural gas used as motor fuel. Fuel. 2019;**239**: 1324-1330

[8] Petryk M, Leclerc S, Canet D, Sergienko I, Deineka V, Fraissard J. Competitive diffusion of gases in a zeolite bed: NMR and slice selection procedure, modelling and parameter identification. The Journal of Physical Chemistry C. ACS (USA). 2015;**119**(47): 26519-26525

[9] Deineka V, Petryk M, Fraissard J. Identifying kinetic parameters of mass transfer in components of multicomponent heterogeneous nanoporous media of a competitive diffusion system. Cybernetics and System Analysis Springer. 2011;**47**(5): 705-723

[10] Ruthven DM. Principles of Adsorption and Adsorption Processes. New York: John Wiley; 1984. 433 p

[11] Kärger J, Ruthven D, Theodorou D. Diffusion in Nanoporous Materials. Hoboken: John Wiley & Sons; 2012. 660 p

[12] Sergienko IV, Deineka VS. Optimal Control of Distributed Systems with Conjugation Conditions. New York: Kluwer Academic Publishers; 2005

[13] Tikhonov AN, Arsenin VY. Solutions of Ill-Posed Problems. Washington D.C.: V.H. Winston; New York: J. Wiley; 1977

[14] Lions J-L. Perturbations Singulières Dans les Problèmes Aux Limites et en Contrôle Optimal. New York: Springer; 2008. Lecture Notes in Math. Ser

[15] Sergienko IV, Petryk MR, Leclerc S, Fraissard J. Highly efficient methods of the identification of competitive diffusion parameters in inhomogeneous media of nanoporous particles. Cybernetics and Systems Analysis Springer. 2015;**51**(4):529-546

[16] Ivanchov M. Inverse Problems for Equations of Parabolic Type. Mathematical Studies. Monograph Series. Vol. 10. Lviv: VNTL Publishers; 2003

[17] Lenyuk M, Petryk M. The Methods of Integral Transforms of Fourier-Bessel with Spectral Parameter in Problems of Mathematical Modeling of the Mass Exchange Process in Heterogeneous Medias. Kyiv: Naukova Dumka; 2000

Chapter 3

Zeolites as Chameleon Biomaterials: Adsorption of Proteins, Enzymes, Foods, Drugs, Human Cells, and Metals on Zeolite Membranes with Versatile Physics-Chemical Properties

Adalgisa Tavolaro, Silvia Catalano and Palmira Tavolaro

Abstract

This chapter is dedicated to demonstrating how both the hydrothermal synthesis of crystalline zeolites with precise atomic compositions and the knowledge of their physics-chemical characteristics allow designing selective materials, useful as powerful tools for biomedical applications. The adsorption of proteins and enzymes, dyes, and drugs and the preparation of scaffolds for in vitro testing of new food and cosmetic formulations are discussed according to the configuration, the composition, and the morphology of prepared materials. Finally, the study of the chemical, molecular, and supramolecular interactions between interesting biological species, drugs, cells, and synthetic materials was used to produce advanced materials and active scaffolds.

Keywords: zeolite membranes, adsorption, proteins and enzymes, drug delivery, cell cultures, food analysis

1. Introduction

The problem of creating excellent biomaterials is a scientific challenge with enormous effects on the economy and health field. Today, synthetic zeolitic materials are probably an opportunity to prepare a successful alternative to traditional biomaterials. Study, chemical design, and manufacture of various inorganic structures allow to have active biomaterials which, in all applications, avoid unwanted responses of the body such as thrombosis [1], inflammatory reactions [2], and infections [3, 4]. In fact, the chemical and physical characteristics of the internal and external surfaces of these microporous materials lend themselves to well interact with exchangeable and reactive ions (e.g., toward drugs and microbes), with proteins (preserving their biological reactivity), and with cells (being non-cytotoxic). Certainly, the chemical approach must be modified by replacing the traditional parameters used to characterize biomaterials with novel concepts such as

IntechOpen

contact angle with point of zero charge (PZC) and wettability with silicon/aluminum ratio. All the most advanced applications concern materials that occur in a membrane configuration [5], i.e., having chemical and physical selectivity whether they are pure materials, in mixture, or made of overlapping layers (composites). Zeolites, already in the form of crystals, have selectivity (shape selectivity, hydrophobicity/hydrophilicity), which can be modified by means of chemical functionalization, ion exchange, impregnation, etc.

2. Traditional biomaterials

A biomaterial has been defined as any substance (other than a drug) or a combination of substances, synthetic or of natural origin, which can be used for any period of time, in whole or in part of a system that treats, increases, or replaces any tissue, organ, or body function.

In 1992 Black defined biomaterial as "a non-living material used in a medical device, designed to interact with biological systems."

The fundamental requirements of every biomaterial are compatibility with human tissues and possessing all those physical, chemical, and biological characteristics that allow the material to adequately perform the task for which it was designed, such as constituting a resistant support, replacing fabrics lost, and promoting regrowth of damaged tissues.

Today, more and more, research in the field of biomaterials is fueled by the need to find new materials that can last a long time, due to the increase in the average life of the population, the increased need for prostheses even by young people, and the need to reduce the number of revisions that weigh on public health costs. Furthermore, the biological materials deriving from homologous or heterologous transplants have shown important problems: limited availability, need for a further surgical operation, potential transmission of infectious diseases, reduced osteoconductive capacity, and limited ability to be incorporated into the host bone.

Biomaterials can be divided into three main types based on the response that they generate in the host tissue: an inert material does not cause a response in the tissue, a bioactive material is integrated by the surrounding tissues, and a degradable material is reabsorbed and incorporated into the surrounding tissue and can even dissolve completely after a certain time. To date, the materials most frequently used in medical applications are metals, typically inert and used for applications subjected to loads, with sufficient fatigue resistance to withstand daily activity; ceramics, used for their hardness and resistance to stress in applications such as joint surfaces, in teeth and in surfaces in contact with bone; and polymers, used for their stability and flexibility but also for low friction in articular surfaces (**Figure 1**).

2.1 Important physics features

2.1.1 Morphology

Substrate morphology can influence cell adhesion, influencing the substrate's ability to adsorb proteins and/or altering the conformation of adsorbed proteins.

For example, material roughness affects the adhesiveness of platelets. Blood normally coagulates when exposed to surfaces different from the biological endothelial ones; for this reason various attempts have been made to find a synthetic material that is biocompatible with blood. When the surfaces were tested in a laminar flow cell, it is noted that an added surface roughness led to a decrease in platelet adhesion on hydrophilic surfaces, while an increase in platelet adhesion was

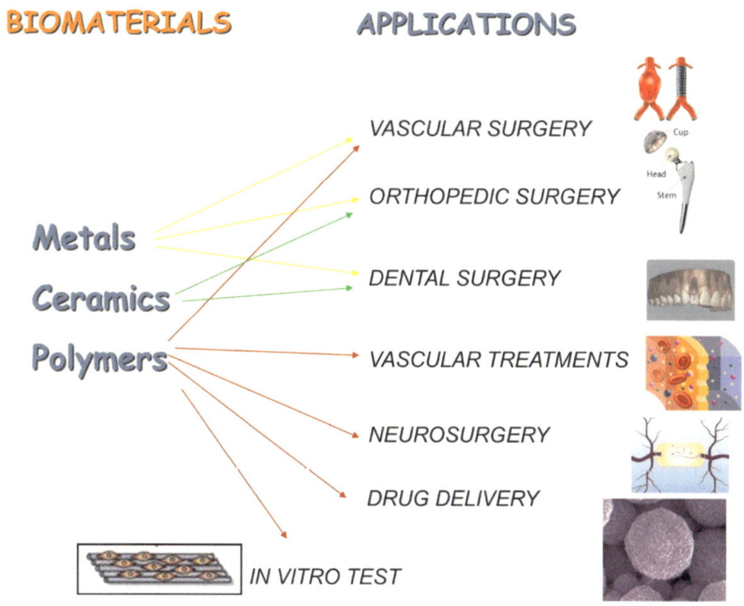

Figure 1.
Type of biomaterials and their biomedical applications.

found on hydrophobic surfaces. This demonstrates that morphology influences the properties of the material and therefore its interaction with the cells.

2.1.2 Topography

Another aspect of the morphology that must be considered to control the cellular response is surface topography. Topography, coupled with biochemical and physical signals, regulates cellular functions such as migration, adhesion, morphogenesis, differentiation, and apoptosis. Topography not only allows the systematic study of cell-substrate interactions but can also control cell orientation and morphology, which in turn controls other cellular responses. Therefore the techniques used to create substrate, precise, high-resolution surfaces acquire great importance. Today, topographies are generated with a resolution up to micrometer level, but with the advancement of modeling techniques and technology, the resolution level will reach the nanometer scale like the most in vivo structures (such as the collagen fibrils of the basement membrane).

As previously stated, topography can induce changes in cell morphology, thus influencing cellular responses such as proliferation, gene expression, and cellular function. These responses also vary depending on the type of cells being used for sowing. For example, the experimentation conducted with surfaces on which channels have been produced revealed that many cell types tend to line up along the main axis of the channels themselves and that often the organization of the cytoskeletal components and the focal contacts is oriented in the same direction. The degree of cellular alignment in the direction identified by the channels depends in a complex way on the characteristics of the topographical surface structure.

Finally, it has been observed that also the symmetry and regularity of the topographical structure are important properties of the substrate that influence cellular behavior. The results showed that regular topography reduces cell adhesion very markedly, while surface discontinuities have improved cell adhesion. This shows that the substrate topography is important for cell adhesion and therefore for cell-substrate interaction.

2.1.3 Stiffness

Stiffness of a material is measured with the modulus of elasticity or Young's modulus. It is important to have sufficient substrate stiffness for the anchor-dependent cells to adhere to the surface. It is fundamental for the characterization of the interactions that modulate intracellular signaling pathways and cellular events, from gene expression to cellular locomotion. In fact, cell movement can be guided by manipulation of substrate stiffness characteristics. It has been shown how the mechanical properties of the matrix influence the differentiation of stem cells. Moreover, the proliferation and cellular mobility varied as the stiffness of the substrate varied. In particular, different types of substrates with different stiffness seeded with NSPC2 cells showed that the optimal stiffness for proliferation was 3.5 kPa, while for neuronal differentiation, it is less than 1 kPa [6].

2.1.4 Crystallinity

By controlling the amorphous-crystalline microstructure of the surface layer of the substrate, it is possible, for example, to improve the compatibility of blood surfaces. Surfaces with different degrees of crystallinity were tested, and an increase in the adhesiveness of the platelets was noticed on substrates that had less crystallinity. The particular amorphous-crystalline surface microstructure also modified the denaturation of adsorbed proteins. For example, the particular amorphous-crystalline microstructure of apolar surfaces such as propylene (with 55% surface layer crystallinity) has been shown to reduce platelet activity [7].

During the design of scaffolds for in vivo implantation, crystallinity can also influence the biodegradability of the scaffold and consequently the cellular response. The crystalline region is in fact more resistant to water infiltration and therefore delays the degradation of the biomaterial. For example, the adhesion, proliferation, and morphology of human chondrocytes of articular cartilage tested as a function of the crystallinity of various degradable polymers. The results suggested that cell proliferation is slower on crystalline polymers than amorphous polymers. This highlights the interesting dynamics between cell and substrate depending on the crystallinity of the material.

A variation in crystallinity can also affect surface roughness, on a nanometric scale. Osteoblasts seeded on polymeric substrates having different crystallinity and their number were measured using fluorescence microscopy. The results showed that the proliferation rate was greater on the smooth regions of the substrates, while it was smaller on the rough regions; a decreasing monotonic variation of proliferation as a function of roughness was observed. The critical roughness above which there is a significant reduction in the proliferation rate is 1.1 nm. It has therefore been shown that the cells respond directly to the topography of the substrate, as they are sensitive to nanometric variations in the substrate topography.

2.2 Important chemical features

2.2.1 Wettability

Wettability of a solid surface is a measure of its hydrophobicity and hydrophilicity. It concerns to the ease liquid phase spreading on a solid surface, which, for polymeric materials, is generally evaluated by contact angle measurements. Contact angle represents the angle formed by the intersection of liquid-solid and liquid-vapor interface, obtained by virtual tangent drawing along the vapor-liquid interface. Water molecules are not able to form hydrogen bonds with the hydrophobic support; therefore, they form hydrogen bonds between them generating a more ordered structure with less entropy. Water molecules on a polymeric surface reorganize around proteins, causing the irreversible unfolding and adsorption of native proteins on the substrate surface. The proteins present in the serum can act as surfactants, or they can lower the surface tension of a liquid; the hydrophobic domains interact with the substrate and the hydrophilic domains form hydrogen bonds with the water molecules, thus facilitating the wettability of the surfaces. This involves a release of ordered water molecules, which is energetically favorable due to the increase in entropy, known as hydrophobic effect. In general, proteins are preferentially adsorbed on hydrophobic surfaces, mediated by their hydrophobic domains. Instead of seeing the underlying surface, the cells see the layer of proteins adsorbed on the surface of the substrate, which then modulate cell adhesion.

2.2.2 Surface charge

Polymeric biomaterial surface charge affects the adsorption and the unfolding of proteins on its surface. Unlike wettability, the driving force for protein unfolding on a charged surface is electrostatic interaction, not hydrophobic interactions. Protein unfolding depends on the net charge that proteins and cells encounter on the surface in the cell culture medium. Many proteins have a net negative surface charge, which promotes their adsorption on a positively charged surface.

2.3 Zeolites as biomaterials

We can imagine that the adaptability of zeolitic materials to interact with different biologically active molecular species and with different environments containing cells makes them (within the vast field of biomaterials) entirely comparable to chameleons (in the animal kingdom). In fact, it can be imagined that just as the complex specialized organization of cells can produce a color change in chameleons (by acting on well-defined physical parameters), the changing complex chemical organization in zeolite framework can interact with proteins, enzymes, cells, and foods (by modifying the preparation methods). The chameleon-like characteristics of the zeolite membranes as biomaterials can be inferred from the various applications reported in the literature and are highlighted in **Figure 2**.

Zeolites used as biomaterials can be distinguished according to the origin: natural, artificial, and synthetic. It should be noted that the crystallized structures by means of hydrothermal reactions in the laboratory under controlled conditions have, at the same time, higher crystallinity, purity, and reproducibility of chemical composition, zeolitic structure, dimensions, morphology, and distribution of the pore and channel system. The choices of chemical parameters, in the synthesis and in the pre- or posttreatments, always have repercussions on the macroscopic chemical-physical characteristics of prepared materials, such as the hydrophobicity, the point of zero charge (PZC), and the presence of the various types of ions, present in the form of exchangeable cations or clusters. We can analyze these different characteristics by gradually shifting our analysis from the microscopic atomic field to the macroscopic membrane field.

2.4 Synthesis and characterization of zeolite biomaterials

Zeolites are bi-functional materials having both Lewis and Brönsted acidity. These two types of acidity are not independent of each other but are closely related to each other, and both participate in the formulation of the total acidity and hydrophilicity of the zeolitic material. Lewis acidity is linked to the presence and relative concentration of trivalent aluminum atoms within the framework (and of other chemically equivalent atoms); therefore, it strongly depends on the so-called SAR ratio. Hydrophobic zeolites are not very acidic (like silicalite-1 or silicalite-2), while hydrophilic zeolites (like zeolite Y or zeolite A) or isomorphically substituted (with cation 3^+) have a high value of Lewis acidity (**Figure 3**).

The second type of acidity is linked to functional silanol groups and more complicated to analyze. We must distinguish between Brönsted acidity of the single bonds (electronically influenced by the neighboring atoms) and the total acidity (depending both on the chemical composition and on the reaction environment used for the synthesis). As a first approximation, we can consider Brönsted total acidity concerning the outer surface of the crystals so it is probably the most important type in interactions with species larger in size than that of zeolite pores such as human cells (**Figure 4**).

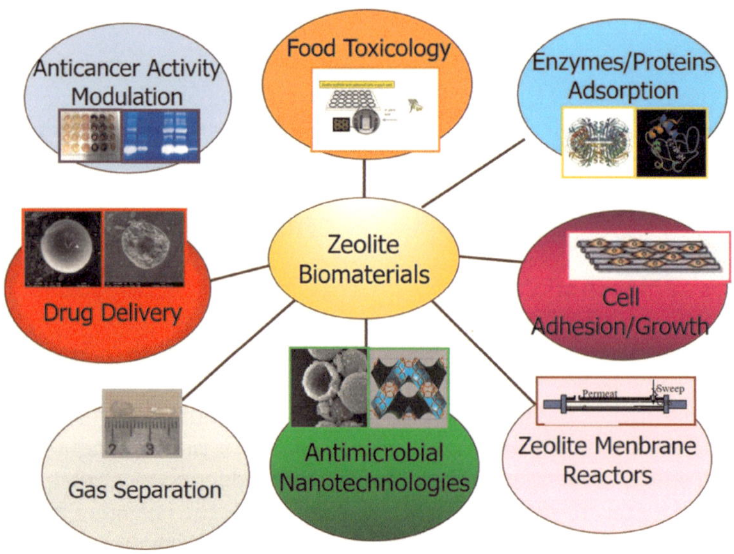

Figure 2.
Application fields of zeolite biomaterials.

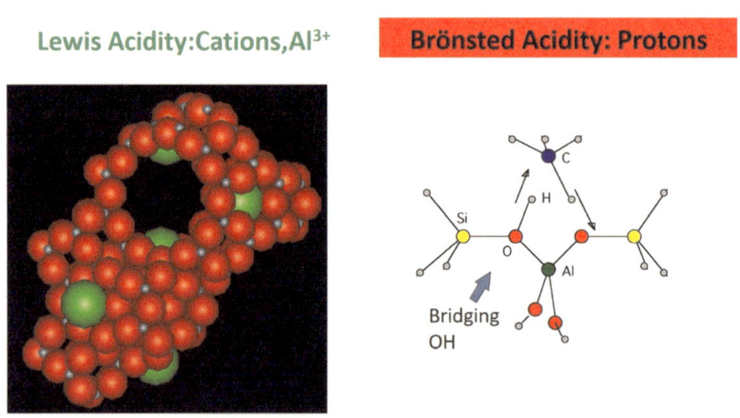

Figure 3.
Scheme of Lewis and Brönsted acidity of single bonds. The substitution of tetravalent Si atoms in the lattice with the trivalent Al atoms generates local negative charges, which are then compensated by extra-framework cations. The charge compensation by protons results in strong Brönsted acid sites.

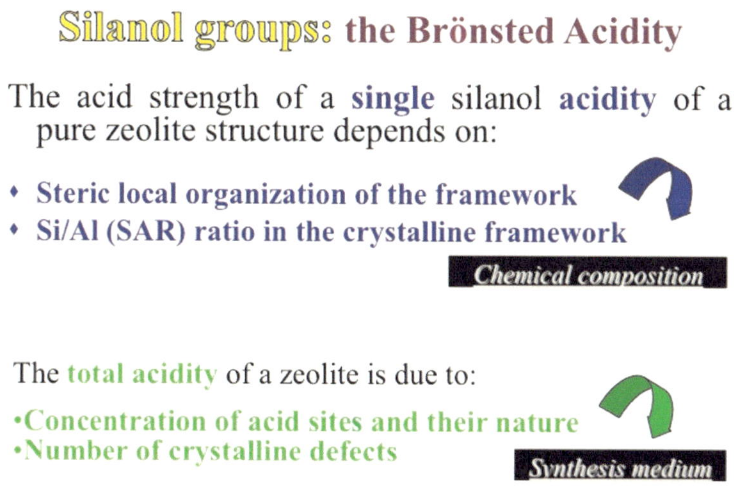

Figure 4.
Schematic summary of the Brönsted acidity for zeolitic biomaterials.

Finally, an extremely important characteristic is point of zero charge (PZC), which is the pH value at which the zeolitic membrane is electrically neutral.

Zeolites have been approved and defined as safe for human consumption by the FDA (Code of Federal Regulations, April 2017). They are also widely used in agriculture as fertilizers because they are declared nontoxic by IARC (IARC, Lyon, Vol. 68, 5061997). Furthermore, they have been approved by the Codex Alimentarius Commission for their use as fertilizer (Codex Alimentarius Commission, 2016) and EFSA experts as flavoring material or food storage adjuvant (EFSA J 11: 3155, 2013) and use as feed additives (EFSA J11: 3039 2013).

Only a natural type of fibrous zeolite is considered dangerous and erionite [8, 9]. It can induce, if breathed, tumors (pleural and peritoneal mesothelioma) and pathologies of the respiratory system, caused by its microfibrous morphological characteristics.

Another use of zeolites, with good results, is to add them as additives for animal feed. The integration of zeolites in animal feed has been studied on different animals (sheep, calves, pigs, etc.), and the results obtained have shown that these substances allow preventing diseases. Moreover, the use of zeolite in animal nutrition improves the assimilation of nutrients and therefore, consequently, promotes an increase in weight of the animal itself [10].

In recent years, zeolitic membranes have been studied as new biomaterials used for biomedical applications. These membranes, in fact, are considered as an ideal support for the immobilization of biological molecules not having the limitations associated with traditional polymeric membranes [11].

Numerous biological molecules were adsorbed and immobilized on zeolite crystals and membranes. These species include cytochrome c [11], bovine serum albumin (BSA) [12], glucose [13], uremic toxins [14, 15], nitrosamines [16], and catalase [17].

An important characteristic of zeolitic membranes is that the basic/acid nature of the material can be modified by varying the Si/Al ratio or by introducing different metals (Me) into the crystalline structure and changing the Si/Me ratio [12].

Cytochrome c is a water-soluble electron carrier that is efficiently immobilized onto zeolitic membranes, but the composition of the membrane is an important factor that influences the immobilization performance. We also studied adsorption of BSA protein [12] on FAU, BEA, and MFI zeolite crystals synthesized under hydrothermal conditions and membranes showing that the chemical composition

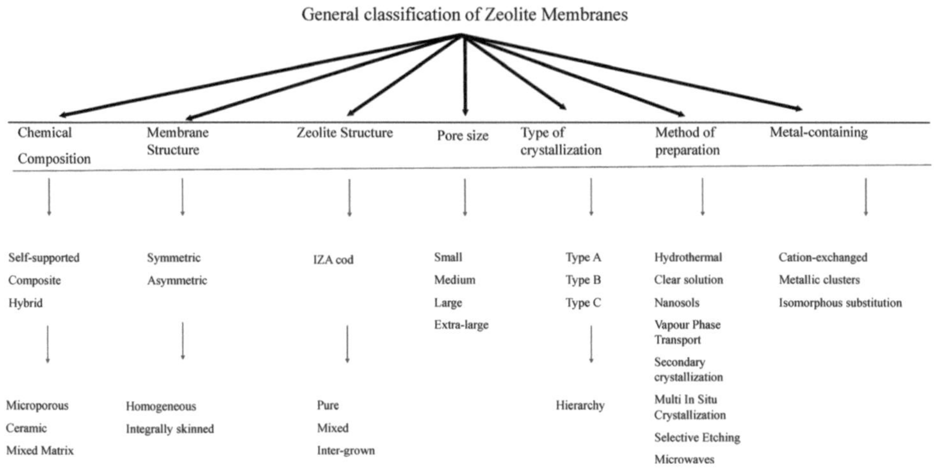

Figure 5.
Classification of zeolite membranes.

and structure of zeolitic membranes influences protein adsorption kinetics. In our work the acidity of zeolite structures was modulated considering several frameworks and MFI structures (having isomorphous vanadium atoms incorporated into the crystalline structure), and Si/Al, Si/V, and Al/V ratios were varied changing the chemical composition of the gels' reaction precursors.

Obtained results shows that zeolite Y surface adsorbs largest amounts of BSA and its percentage of adsorption increases with temperature and depends on the pH of the solution used, being absolute maximum in correspondence of protein pI value. The adsorption difference between the various types of zeolite also depends on the type of hydrothermal crystallization within the inorganic support [12].

2.5 Classification of zeolite membranes

Zeolitic membranes are membranes in which the selectivity is due to the zeolitic structure regardless of the membrane constitution and morphology (**Figure 5**). Composite, self-supported, mesoporous, and mixed-matrix membranes can certainly be considered zeolitic if the chemical-physical zeolite characteristics influence the process to which they are applied.

Zeolitic membranes consist of intergrown crystals of sizes in the range of a few nanometers up to several hundred microns. They can be formed in self-supported zeolite membranes, which are very fragile, and therefore, for applications that require the use of pressure gradients, microporous film is grown using permanent inorganic (e.g., ceramic and metal) and organic (e.g., plastic and wood) supports.

Zeolitic membranes can be classified into various categories based on their shared physics-chemical and morphological characteristics, using various classification parameters. Currently, we can identify eight possible classification systems that can coexist and are preferred by scientists depending, for example, on the material application or the feature studied. These systems can thus be identified on the inclusion criterion used:

- Chemical composition
- Membrane morphology
- Type of crystalline zeolite framework

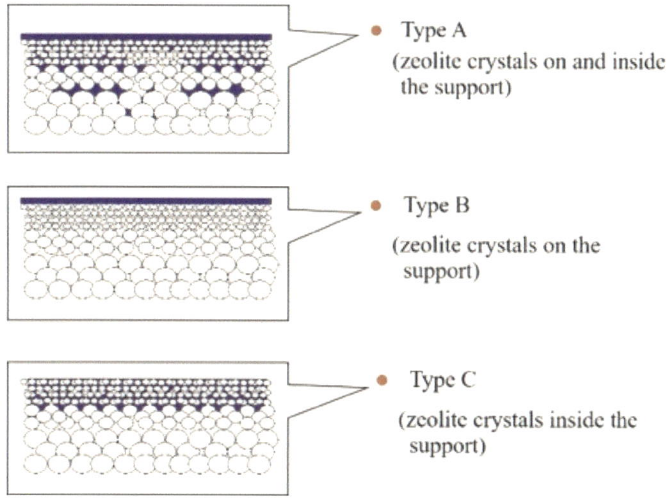

Figure 6.
Schematic representation of different types of zeolite membrane crystallizations.

- Pore size
- Type of crystallization (**Figure 6**)
- Synthesis methodology
- Metal-containing frameworks
- Origin (natural, artificial, synthetic)

Classification based on the membrane chemical composition includes inorganic, composite, and hybrid membranes. Inorganic zeolitic membranes are solid membranes made of zeolitic crystals and/or inorganic materials such as metals, oxides, amorphous silica, and ceramic particles. These membranes are very stable and resistant to mechanical stress at high temperatures and pressure gradients.

Composite membranes are constituted by a superposition of different materials layers, which can be evidenced by an orthogonal section to the surface like zeolite/alumina membranes.

Hybrid membranes (mixed-matrix membranes) include membranes consisting of zeolitic crystals dispersed in a polymeric film. These membranes have great tensile strength and great elasticity but low thermal and mechanical resistance as well as poor aging stability. They are prepared easily and quickly by means of inclusions or crystal depositions.

3. Adsorption

The possibility of modulating the specific characteristics of zeolites in a membrane configuration using inorganic zeolitic membranes in biotechnological applications such as molecular separations, enzymatic membrane reactors, protein chips, drug delivery, etc. is an attractive perspective that would offer remarkable potential applications. Naturally, the selection of materials suitable for specific applications cannot ignore the study of the interaction between the biological species and the crystalline inorganic support and therefore the understanding of adsorption.

Although the zeolitic materials have been well characterized and widely used in chromatographic applications, the analysis of protein adsorption on zeolitic crystals is poorly reported in the literature, and even less numerous are the research activities concerning the zeolitic membranes.

Adsorption is a surface phenomenon characterized by the interaction of a molecular species present in a solution (adsorbate) with the external or internal porous surface of a solid (absorbent). In the thermodynamic sense, most studies have considered adsorption as a reaction, which, of course, is more extensive if the solid material has a high surface area Eq. (1):

$$A + B_{solid} \leftrightarrow AB \tag{1}$$

The species A dissolved in the solution reacts with the adsorbent B to form AB.

In a thermodynamic equilibrium situation, the Gibbs free energy change tends to zero, whereas the two chemical potentials are equal according to Eq. (2):

$$\Delta G = \mu_{s-l} - \mu_l = \Delta G^0 + RTln(K_e) = 0 \tag{2}$$

where ΔG is the Gibbs free energy change, $\mu_{s\text{-}l}$ is the chemical potential in the solid–liquid interface, μ_l is the chemical potential in the liquid phase, R is the universal gas constant, and K_e is the equilibrium thermodynamic constant (Cheng and Zhang, 2014 da Bonilla):

$$\ln(K_e) = -\frac{\Delta H^0}{RT} + \frac{\Delta S^0}{R} = 0 \tag{3}$$

Eq. (3) permits to calculate the adsorption thermodynamic values of ΔH° and ΔS° plotting $ln(K_e)$ versus I/T values in the van't Hoff plot. A reasonable physic meaning of K_e can be given considering:

$$K_e = \frac{activity\ of\ occupied\ sites}{(\text{activity of empty sites})\ (\text{activity of adsorbate in solution})} \tag{4}$$

$$K_e = \frac{\frac{q_e}{q_m}}{\left(1 - \frac{q_e}{q_m}\right)\frac{C_e}{C^\circ}} \tag{5}$$

and

$$q_e = \frac{q_m K_e\left(\frac{C_e}{C^\circ}\right)}{1 + K_e\left(\frac{C_e}{C^\circ}\right)} \tag{6}$$

This last equation allows to obtain the dimensionless value of K_e by plotting the experimental data obtained for q_e (expressed in moles/grams) versus C_e (expressed in moles per liter) and considering the value of C° equal to 1 mole per liter.

4. Adsorption of proteins and enzymes

It is known that protein molecules selectively bind to non-biological surfaces such as those of the metals, of carbonate oxides, and semiconductors. Naturally, in order to use these inorganic supports as biomaterials, it is necessary that the protein biological activity is preserved with immobilization. In fact, it is possible that the interaction of the inorganic matrix with the protein causes its inactivity or functional slowdown as a consequence of structural or conformational changes or steric unavailability of the active site. Therefore, it is evident that a suitable selection of the matrices is essential to obtain immobilized and, at the same time, active biological species. Zeolites have a large surface area and thermal, mechanical, and chemical resistance; therefore, they are well suited to the role of adsorbent supports for biological molecules. The acid/base nature of the material can be modified changing the silicon/aluminum ratio (called SAR) of the framework or introducing different metal atoms into it (creating isomorphic substitution) and varying the silicon/metal ratio by synthesis. Furthermore, it is possible to modify the acidity of zeolites by ionic exchange of the extra-framework cations present in the microporous channels, for example, with protons.

5. Drug delivering zeolite biomaterials

Most biomaterials used for implants are inert, non-immunogenic, and nontoxic, but devices made with such materials often contain parts that trigger the so-called foreign body reaction, a material rejection complex process still not completely understood and probably related to the presence of histamine and the fibrinogen adsorption onto the implant surfaces [18]. These reactions can produce thrombosis, infections, inflammations [19], formation of fibrotic tissues around implantation, and prostheses. To realize novel active drug-releasing biomaterials, we prepared low-cost, specific drug carrier membranes for innovative biomedical drug delivering materials for implants [20]. In order to achieve this purpose, we synthesized MOR and MFI zeolite nanocrystals and composite membranes using porous stainless steel permanent supports; then we prepared ion-exchanged structures Cu(II) and Zn(II)-containing hydrophilic frameworks.

Our work on the adsorption, and the subsequent release, of a model drug revealed that these zeolite materials are useful to immobilize famotidine (3-[[[2-diathiazolyl]methyl]thio]sulfamoylpropionamidine), a histamine H_2 receptor. Furthermore, we evidenced that the synthesized materials, having different types of zeolitic structure and bivalent counter-cations, show different performances suitable to biomedical applications. In fact, the adsorption percentage of on transition metal-containing nanocrystalline zeolites was greater with respect to the as-made materials suggesting that these cations chemically interact with the drug and that cupric ions form stable organometallic complexes.

6. Interaction of zeolite materials and cells

The composition of traditional scaffolds has changed considerably since the end of the 1980s, when the field of tissue engineering was started in a systematic way. This improvement reflects the greater scientific understanding of the needs of the cells in the adhesion and management of their behavior, which are fundamental in tissue engineering applications. The success of a new scaffold is not only based on its mechanical characteristics or on the surrounding chemical environment but also on its detailed chemical surface and topography (in a nanometer scale). These last two characteristics are not so easy to achieve by chemical synthesis for a large number of inorganic or polymeric materials.

The analysis of cell-substrate interaction is of fundamental importance in order to design biomimetic scaffolds capable of replacing damaged vital organs, or tissues, or to assist the body's natural healing processes. The ability of a cell to recognize and interact with the substrate represents the first indispensable step, without which processes such as cell adhesion proliferation, migration, and differentiation would not be possible. Therefore the understanding of the mechanisms that determine the early phases of cell-material adhesion, as well as their control, is indispensable for the design of biomaterials. Both the mechanical and biochemical properties of the material determine the efficacy and agreed with which the cells recognize the material.

The possibility of modifying and controlling surface properties at the micro-/nanolevel constitutes one of the major breakthroughs, because it opens a whole new range of strategies seeking the desired interaction with the biological environment. In order to prepare a new generation of biomaterials with enhanced properties, a different approach needs to be reached, based on a more fundamental understanding of the way in which the structure of a biomaterial controls its biological activity. The chemical properties influence the surface properties of a material and, consequently, cell behavior. When cells are exposed to a suitable scaffold, a layer of proteins is adsorbed on the scaffold surface within a few milliseconds. Thus cells "see" the layer of adsorbed proteins rather than the actual abiotic surface. The chemistry of the surface of a scaffold can be developed in order to control the adsorption of proteins, which in turn controls cell adhesion. According to the hoped-for result, the chemical characteristics of the surface of a material can be modified to modulate the interactions of cells adherent to the substrate, with consequent influence on morphology, migration, differentiation, proliferation, and cell apoptosis. The effect on cell behavior starts at the point of interaction. Furthermore, the conformation of the surface chemistry also affects the way proteins are immobilized and the adsorption of these on the surface. Starting from this assumption, we designed and prepared various crystalline zeolite scaffolds, which are different depending on the preparation method. It is evident that a porous, crystalline material having an inorganic framework with modulable acidity, hydrophilicity, and pore size constitutes a stable, homogeneous, ion- and solvent-available support. Zeolite membranes symbolize this novel type of chameleonic scaffold.

7. Zeolite scaffolds

Zeolite scaffolds provide a framework above and within which cells seeded in the culture medium can adhere and over time populate them. These processes require that the scaffold structure must also be able to support a growing number of cells allowing the transport of sufficient amounts of nutrients and the removal of waste products having an extended surface where the cells can adhere and migrate freely so as to form a mass of cells with subsequent deposition of an active extracellular matrix (ECM) [21].

A scaffold is a critical component of tissue engineering, as it is intended to release, contain, and form new tissue in vitro or to promote tissue repair in vivo. Porosity, architecture, and rate of degradation are important aspects of the material that allow the growth of cells that guide the formation of tissues such as bone.

An ideal scaffold should have the following characteristics:

- Biocompatibility suitable not to induce any adverse reaction
- Degradation rate appropriate to match the tissue regeneration process
- Narrow pore size distribution to allow cells to have a sufficient space to grow and access to nutrients and metabolites

DOI: http://dx.doi.org/10.5772/intechopen.88422

- High surface area to boost the cell adhesion
- Chemical composition appropriate to promote cell differentiation and growth
- Structural parameters appropriate to modulate cellular biosynthesis

When the adhered cells increase in number, they begin to enter the internal pores of the scaffold. If the porosity and interconnection between the pores are good, the cells grow and colonize the entire scaffold releasing their extracellular matrix. The upper layer of cells consumes more oxygen and nutrients, thus limiting the amount available for the cells that are migrating into the scaffold; the maximum depth at which cells can survive corresponds to the depth of cellular penetration. We studied many types of both self-supported and hybrid PLA-containing zeolitic membranes (MMMs) to study interactions with different types of normal [22] or carcinogenic (MDA-MB-231 [23] and MCF-7 [24]) cells. Initial cell tethering and filopodia exploration are followed by lamellipodia ruffling, membrane activity, and cell spreading. With time endogenous matrix is secreted by the cells, and matrix assembly sites form on the ventral plasma biological membrane. Later, with increased integrin recruitment, these early cell-matrix contacts form anchoring focal complexes at the lamellipodium leading edge that are reinforced intracellularly to form larger focal adhesion plaques upon increased intracellular and/or extracellular tension. The regulation of focal adhesion formation in adherent cells is highly complex and involves both the turnover of single integrins and the reinforcement of the adhesion plaque by protein recruitment. It follows that focal adhesions emerge as diverse protein networks that provide structural integrity and dynamically link the ECM to intracellular actin filaments, directly facilitating cell migration and spreading through continuous regulation and turnover. Furthermore, in combination with growth factor receptors, these adhesive clusters initiate signaling pathways and regulate the activity of nuclear transcription factors and processes crucial to cell growth and differentiation. The adhesion sites act as mechanosensors that form additional contact points with the underlying substratum in response. Preceding focal adhesion reinforcement, a tightly regulated series of temporospatial events occurs, mediating integrin clustering in an anisotropic manner in the direction of force. Our works underlined that the cells of both lines assume a specific morphology under the influence on the major peculiarities of scaffolds.

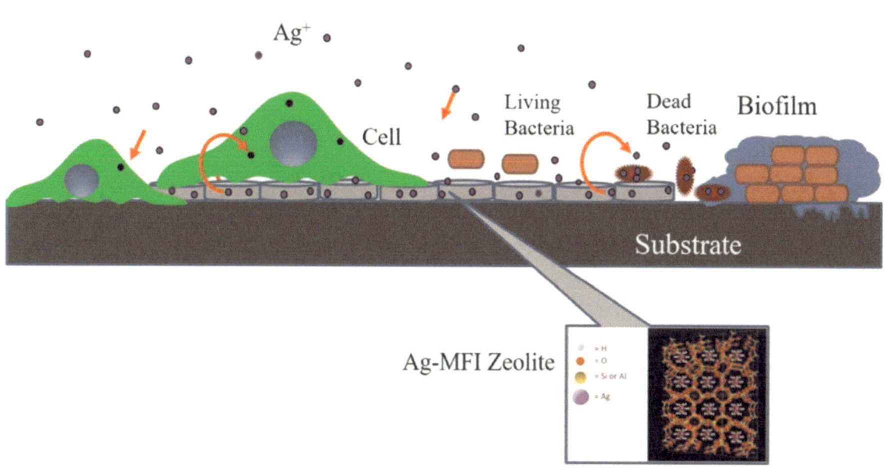

Figure 7.
Schematic representation of the antimicrobial activity of zeolite scaffolds.

Synthetic zeolite scaffolds have been successfully applied to in vitro studies regarding both adhesion and cell growth kinetics [25] as well as to the analysis of new formulation cosmetics and foods [26] (**Figure 7**).

8. Conclusions

For years, zeolite crystals have been used in various technological fields of great industrial interest such as catalysts, ion exchangers, and adsorbents in chromatographic applications. Today, the preparation of crystalline zeolite membranes plays a central role in many new applicative fields such as catalytic zeolite membrane reactors, gas separations, drug delivery, anticancer activity modulation, food toxicology, enzyme/protein adsorption, and antimicrobial nanotechnologies. Zeolite crystals and membranes are key materials for selective adsorption of biological molecules, drugs, and chemotherapy drugs. Moreover, the zeolite membranes represent a synthetic scaffold suitable, ideal, and able to guarantee the survival, growth, and correct differentiation of human normal and cancer cells. The development of zeolite membranes with versatile physics-chemical properties may constitute the goal for new components in biomedical and biotechnological applications.

Acknowledgements

The authors acknowledge Ministero dell'Istruzione, dell'Università e della Ricerca (MIUR) for funding as part of "Spread Bio-Oil PON01_01 00293 project, PON R&C (Programma Operativo Nazionale Ricerca e Competitività 2007-2013)."

Conflict of interest

The authors declare no conflict of interest.

Nomenclature

In this work, the authors use framework type codes of the IZA Commission.

Abbreviations

PZC	point of zero charge
$NSPC_2$	neural stem/progenitor cells
SAR	silicon/aluminum ratio
FDA	Food and Drug Administration
IARC	International Agency for Research on Cancer
EFSA	Education and Skills Funding Agency
BSA	bovine serum albumin
FAU	faujasite structure
BEA	beta-structure
MFI	MFI structure
IZA	International Zeolite Association
ECM	extracellular matrix

PLA	polylactic acid polymer
MOR	mordenite structure
MMM	mixed-matrix membrane
MDA-MB-231	human breast adenocarcinoma cells
MCF-7	human breast ductal carcinoma cells

Author details

Adalgisa Tavolaro[1*†], Silvia Catalano[2†] and Palmira Tavolaro[2†]

1 Research Institute on Membrane Technology, National Research Council of Italy, ITM-CNR, University of Calabria, Cs, Italy

2 Department of Physics, University of Calabria, Cs, Italy

3 Department of Pharmacy, Health and Nutritional Sciences, University of Calabria, Cs, Italy

*Address all correspondence to: a.tavolaro@itm.cnr.it

† These authors contributed equally.

References

[1] Jamiolkowski MA, Pedersen DD, Wu W-T, et al. Visualization and analysis of biomaterial-centered thrombus formation within a defined crevice under flow. Biomaterials. 2016;**96**:72-83

[2] Nilsson B, Ekdahl KN, Mollnes TE, et al. The role of complement in biomaterial-induced inflammation. Molecular Immunology. 2007;**44**:82-94

[3] Raphel J, Holodniy M, Goodman SB, et al. Multifunctional coatings to simultaneously promote osseointegration and prevent infection of orthopaedic implants. Biomaterials. 2016;**84**:301-314

[4] Avani F, Damoogh S, Mottaghitalab F, et al. Vancomycin loaded halloysite nanotubes embedded in silk fibroin hydrogel applicable for bone tissue engineering. International Journal of Polymeric Materials and Polymeric Biomaterials. 2019:1-12

[5] Ishihara K. Bioinspired phospholipid polymer biomaterials for making high performance artificial organs. Science and Technology of Advanced Materials. 2000;**1**:131-138

[6] Leipzig ND, Shoichet MS. The effect of substrate stiffness on adult neural stem cell behavior. Biomaterials. 2009; **30**:6867-6878

[7] Kawamoto N, Mori H, Terano M, et al. Blood compatibility of polypropylene surfaces in relation to the crystalline-amorphous microstructure. Journal of Biomaterials Science. Polymer Edition. 1997;**8**:859-877

[8] Baris B, Demir AU, Shehu V, et al. Environmental fibrous zeolite (erionite) exposure and malignant tumors other than mesothelioma. Journal of Environmental Pathology, Toxicology and Oncology. 1996;**15**:183-189

[9] Gualtieri AF, Gandolfi NB, Pollastri S, et al. Where is iron in erionite? A multidisciplinary study on fibrous erionite-Na from Jersey (Nevada, USA). Scientific Reports. 2016;**6**:37981

[10] Eroglu N, Emekci M, Athanassiou CG. Applications of natural zeolites on agriculture and food production. Journal of the Science of Food and Agriculture. 2017;**97**:3487-3499

[11] Tavolaro P, Tavolaro A. Zeolitic inorganic supports for the cytochrome c immobilization: Comparative study of several zeolite membranes. Desalination. 2006;**200**:516-517. DOI: 10.1016/j.desal.2006.03.418

[12] Tavolaro A, Tavolaro P, Drioli E. Zeolite inorganic supports for BSA immobilization: Comparative study of several zeolite crystals and composite membranes. Colloids Surfaces B Biointerfaces. 2007;**55**:67-76

[13] Wu Q, Niu M, Chen X, et al. Biocompatible and biodegradable zeolitic imidazolate framework/ polydopamine nanocarriers for dual stimulus triggered tumor thermo-chemotherapy. Biomaterials. 2018;**162**: 132-143

[14] Namekawa K, Tokoro Schreiber M, Aoyagi T, et al. Fabrication of zeolite-polymer composite nanofibers for removal of uremic toxins from kidney failure patients. Biomaterials Science. 2014;**2**:674-679

[15] Bergé-Lefranc D, Pizzala H, Paillaud JL, et al. Adsorption of small uremic toxin molecules on MFI type zeolites from aqueous solution. Adsorption. 2008;**14**:377-387

[16] Zhou CF, Zhu JH. Adsorption of nitrosamines in acidic solution by zeolites. Chemosphere. 2005;**58**:109-114

[17] Tavolaro A, Catalano S, Tavolaro P. Zeolite membranes to immobilize catalase. Inter. J. Environmental & Agriculture Research (IJOEAR). 2018;**4**: 37-41. ISSN: 2454-1850. DOI: 10.5281/zenodo.1238790

[18] Zdolsek J, Eaton JW, Tang L. Histamine release and fibrinogen adsorption mediate acute inflammatory responses to biomaterial implants in humans. Journal of Translational Medicine. 2007;**5**:31

[19] Tang L, Eaton JW. Inflammatory responses to biomaterials. American Journal of Clinical Pathology. 1995;**103**: 466-471

[20] Tavolaro A, Riccio II, Tavolaro P. Hydrothermal synthesis of zeolite composite membranes and crystals as potential vectors for drug-delivering biomaterials. Microporous and Mesoporous Materials. 2013;**167**:62-70

[21] Hutmacher DW. Scaffold design and fabrication technologies for engineering tissues: State of the art and future perspectives. Journal of Biomaterials Science. Polymer Edition. 2001;**12**:107-124

[22] Tavolaro P, Catalano S, Martino G, et al. Zeolite inorganic scaffolds for novel biomedical application: Effect of physicochemical characteristic of zeolite membranes on cell adhesion and viability. Applied Surface Science. 2016; **380**:135-140

[23] Tavolaro P, Martino G, Andò S, et al. Zeolite scaffolds for cultures of human breast cancer cells. Part II: Effect of pure and hybrid zeolite membranes on neoplastic and metastatic activity control. Materials Science and Engineering: C. 2016;**68**: 474-481

[24] Tavolaro P, Catalano S, Tavolaro A. Effect of physicochemical characteristics of pure and hybrid zeolite scaffolds on neoplastic cell activities. Advanced Science Letters. 2017;**23**:5844-5846

[25] Tavolaro P, Martino G, Andò S, et al. Fabrication and evaluation of novel zeolite membranes to control the neoplastic activity and anti-tumoral drug treatments in human breast cancer cells. Part 1: Synthesis and characterization of pure zeolite membranes and mixed matrix membranes for adhesion. Materials Science and Engineering: C. 2016;**69**: 894-904

[26] Tavolaro P, Catalano S, Tavolaro A. Anticancer activity modulation of an innovative solid formulation of extra virgin olive oil by cultured zeolite scaffolds. Food and Chemical Toxicology. 2019;**124**:139-150

Chapter 4

Anomalous Diffusivity in Porous Solids: Levitation Effect

Shubhadeep Nag and Yashonath Subramanian

Abstract

Fluids confined to zeolites and other porous solids exhibit many distinct properties. One such property is the diffusivity, which exhibits anomalous dependence on the size of the guest molecule confined to the pore. This is termed the levitation effect. A diffusivity maximum as a function of the diameter of the guest is seen. The diameter for which the guest has maximum diffusivity is seen to be associated with a minimum in the activation energy. The existence of similar behavior in other porous solids, framework flexibility, and effect of temperature are discussed. Experimental verification of the existence of the anomalous maximum is then discussed. Diffusion of *n*-hexane and its isomers in zeolite NaY are then discussed in detail. The reduction in the end-to-end length of *n*-hexane while passing through the 12-ring window and the reasons for the same are discussed. Section 3 discusses possible observations of size-dependent maximum in other condensed matter systems.

Keywords: diffusivity, zeolites, porous solids, anomalous diffusivity, size dependence

1. Introduction

Zeolites are excellent catalysts due to the presence of acid sites on the framework. These catalyze a number of reactions such as hydrocarbon isomerization, hydrocarbon cracking or transformation. In recent times, there have been considerable increase in investigations pertaining to various aspects of zeolites [1–13]. There have been several studies aimed at clarifying the mechanism involved in the reactions catalyzed by zeolites. Boronat and Corma have discussed the changes in energy on proton exchange between zeolite and an adsorbate. They attempt to compute the contribution due to the van der Waals interaction between the zeolite and the adsorbate, which can be significant when the molecules are of size similar to the pore [1]. Davis and coworkers have attempted synthesis of enantiomerically enriched molecular sieve. Recently they succeeded in such a synthesis and this has been discussed in a recent article [11, 13]. Corma and coworkers have discussed various novel approaches to the synthesis of zeolites [3]. Dusselier and Davis have discussed at length the synthesis and the use in catalysis of small pore zeolites [12]. Corma and coworkers have discussed the synthesis of a new all-silica polymorph ITQ-55, which is highly efficient in the separation of ethane and ethylene with a high selectivity of 100 [4]. Prashant Kumar et al. [9] have pointed out the novel synthesis of MFI nanosheets with sandwiched MEL. Zeolite can be used for separation of xylene isomers with a high degree of separation [8]. From these it is evident that many new and novel aspects of zeolites are still being discovered and zeolites continues to be an exciting field of research with a bounty of surprises.

Zeolites are porous aluminosilicates capable of accommodating molecules within the pores. They are well known for their catalytic, ion-exchange, and separation properties. They are widely used in petrochemical industries for processing hydrocarbons. Hydrocarbon cracking, transformation, isomerization, etc. are achieved with the help of zeolites [14, 15]. Zeolites are also used in separating hydrocarbon molecules of various sizes [16]. Larger hydrocarbons such as C15 and with still higher number of carbon atoms diffuse slowly through the pores and hence reach the bottom of a zeolite column last. Small molecules such as C1-C5 diffuse fast and exit from the column first [17, 18]. Other molecules of intermediate size have values of diffusivity in between those of C15 and C1-C5 and exit at intermediate times. Thus the various fractions from crude can be separated. This separation is much more energy efficient as compared to separation by distillation.

Another application of zeolites is its use for ion-exchange and water softening [19, 20]. Ions such as Ca^{2+} can be exchanged with Na^{+} already present in the zeolites, thus removing these ions from water. Zeolites are therefore used in detergents for washing clothes.

Apart from the use of zeolite for separation based on the size of the molecule, it is also of considerable use for separations based on the shape of the molecules. This property of zeolites is often referred to as shape selectivity [21]. The pore dimensions of zeolites, which are not always of regular shape, make this possible. An oft quoted example is the separation of xylenes (p-, m-, and o-xylenes) using silicate or ZSM-5 [22, 23]. At sufficiently high temperatures, o-xylene will convert to p-xylene and ZSM-5 also acts as a catalyst.

Zeolites and other host materials also exhibit interesting properties. They exhibit window effect, single file diffusion, and levitation effect [24–26]. Here we will focus on the levitation effect, which refers to the dependence of diffusivity on the guest or diffusant diameter.

2. Diffusivity of guest molecules confined to the pores from MD simulations

In order to understand the diffusion of hydrocarbon and other molecules within the confined spaces of the zeolite, it is essential to carry out an investigation into diffusivity of guest molecules within the confined space provided by a zeolite or other porous solids. Such a study is essential for an understanding of the process of separation of hydrocarbon as well as other mixtures.

One of the early studies investigating the diffusivity of guest molecules in zeolites was the diffusion of xenon in zeolite Y and argon in NaCaA [27]. The diffusivity is likely to be strongly influenced by the bottleneck for diffusion. In the case of xenon the bottleneck is the 12-ring window, which has a diameter of around 7.8 Å. In the case of argon the bottleneck is an 8-ring window, which has a diameter of around 4.5 Å. The ratio of the bottleneck to the molecular/atomic diameter for xenon-NaY system is 7.0/4.1 = 1.70 while in the case of argon-NaCaA it is 4.0/3.405 = 1.17. Although the diameters of the windows are approximate, it is clear that the window diameter is significantly larger than the diameter of xenon while this is not true for argon where the diameter of the window is only slightly larger than the argon diameter. From these it is evident that xenon in NaY should have a higher diffusvity than argon in NaCaA.

From MD, it was found that xenon in NaY has a diffusivity of 0.19×10^{-8} m^2/s while argon in NaCaA has a diffusivity of 0.9×10^{-8} m^2/s. As the diffusivity of argon in NaCaA is higher than xenon in NaY, further investigations were carried out to find the reasons for this. To start with the energy barrier at the bottleneck was computed. This is shown in **Figure 1** [27]. From the figure it is seen that the energy barrier for xenon at the window in Y is positive while the barrier at the window is negative for argon in A zeolite. This explains why the diffusivity of argon in A zeolite is higher than xenon in Y zeolite. The trends seen in the observed barrier appears to be due to the strong interaction of argon with the oxygens of the 8-ring window. As argon is about the same diameter as the window, its strength of interaction with the oxygens is optimum being close to ϵ, which occurs at a distance at which the Lennard-Jones curve is minimum in energy. This is not the case for xenon in zeolite Y where xenon can be close to only some of the oxygens of the 12-ring window. This is the first indication that nongeometrical factors can influence the diffusivity. This study shows that sorbate-zeolite interaction plays an important role.

This study suggests that an understanding of diffusivity as a function of the diameter of the guest species might show something interesting. Such a study was carried out and the results were indeed found to be interesting [28]. A molecular dynamics study of monatomic guest molecules confined to zeolite NaY and NaCaA were carried out in which the diameter of the guest molecule was varied. The diffusivities of the guest species was computed from the time evolution of the mean square displacements. A plot of diffusivity as a function of the reciprocal of square of the guest diameter is shown in **Figure 2** for guests in both zeolites Y and A [28]. It is seen that the diffusivities decrease linearly with increase in the reciprocal of the square of the diameter of the guest molecule for small diameters. This is referred to as the linear regime (LR). As the diameter increases, it is seen that the diffusivity suddenly increases and later decreases sharply exhibiting a maximum in diffusivity. This is referred to as the anomalous regime (AR). This increase followed by a decrease in diffusivity was surprising and needed further investigations.

As can be seen the location of the guest diameter at which the maximum occurs is different in both zeolite Y and A (see **Figure 3** [28]). In order to understand the reasons for the maximum in diffusivity we have tried to search in literature any report that refers to such an observation. Derouane and coworkers have reported a finding arrived at through a theoretical analysis. They showed that the nesting

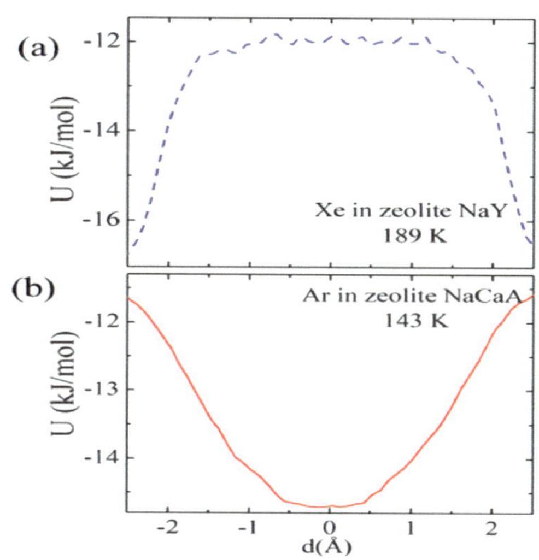

Figure 1.
Potential energy landscape of (a) xenon in zeolite NaY at 190 K and (b) argon in zeolite NaCaA at 140 K. the energy landscapes are computed from molecular dynamics simulations.

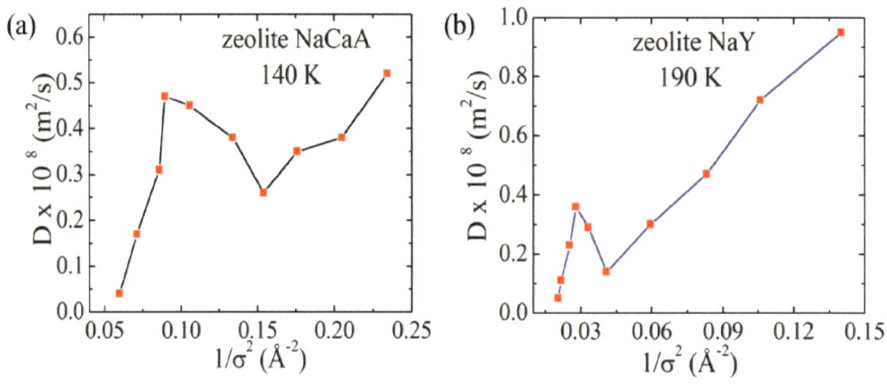

Figure 2.
Diffusion coefficient of guest particles is plotted as a function of $1/\sigma^2$, where σ is the guest diameter. D for guest confined to (a) zeolite NaCaA at 140 K and to (b) zeolite NaY at 190 K temperature, where $1/\sigma^2$ is the inverse square of the vdw radius of guest particles.

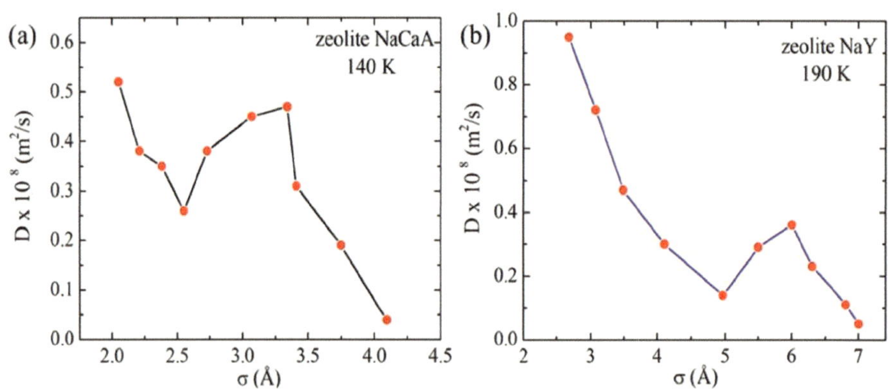

Figure 3.
Diffusion coefficient, D as a function of vdw radius of guest particles, σ is plotted in (a) zeolite NaCaA at 140 K and in (b) zeolite NaY at 190 K temperature.

effect can lead to *floating molecules* when the pore diameter is comparable to the molecular diameter. Earlier Kemball found that when guest molecules are sorbed inside host materials such as zeolites or other adsorbents it is seen that some undergo little loss of entropy. In such systems, he suggested, the guest molecules will exhibit high mobility or *superdiffusivity*.

In order to obtain a better understanding, we define a dimensionless parameter

$$\gamma = \frac{2 \times 2^{1/6}\sigma_{gz}}{\sigma_w} \quad (1)$$

where the numerator gives the distance at which the interactions between the guest and the zeolite atoms are optimum, that is, when this gives an interaction energy of ϵ_{gz}. The denominator is the window radius. Thus, the dimensionless parameter suggests that when these two are equal ($\gamma = 1$) and when they are not equal ($\gamma << 1$). We now plot the diffusivity as a function of γ. This is shown in **Figure 4** [28]. We see that the maximum is seen when γ is between 0.9 and 1.0 for *both* the zeolites. Thus, the guest-zeolite interaction is optimum at $2^{1/6}\sigma_{gz}$ and when this equals $\sigma_w/2$, the diffusivity maximum is seen. This situation when γ is close to unity is illustrated in **Figure 5** along with the situation when $\gamma << 1$. Now when γ is close to unity the guest molecule is passing through the center of the window. Such a position has inversion symmetry, which leads to mutual cancelation of forces

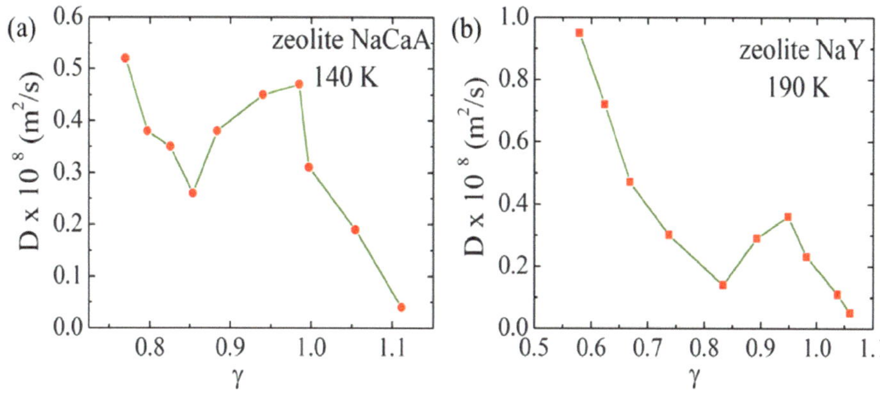

Figure 4.
Diffusion coefficient, D is plotted as a function of levitation parameter, γ (see text) in (a) zeolite NaCaA at 140 K and in (b) zeolite NaY at 190 K temperature.

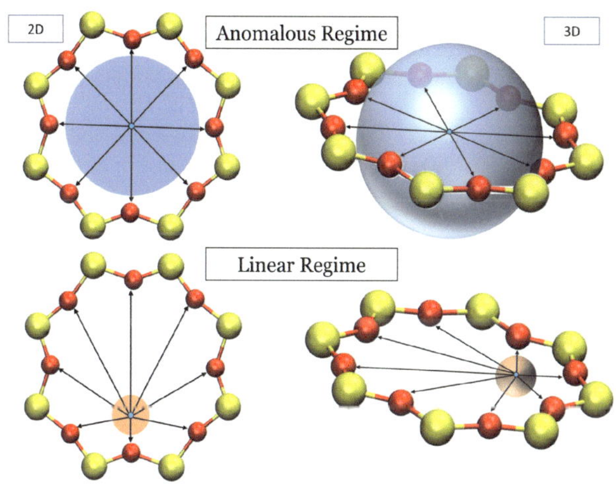

Figure 5.
Schematic figure indicating the position of bigger and smaller particles while passing through the zeolite window. It is shown that the bigger particle passes through the symmetry position, the center of the window, whereas the smaller particle is near to the periphery of the window. Therefore, the forces along a given direction is equal and opposite to that exerted on the bigger particle from the diagonally opposite direction, resulting in higher diffusion than for the smaller particle, which is attracted to the periphery, and therefore experiences a net attraction. This is shown in both 2D and 3D.

leading to rather small force on the guest due to the zeolite. This situation is akin to the guest being a free particle even when confined within the zeolite and therefore has a high diffusivity.

When $\gamma << 1$ the guest passes through the window at the periphery. These points do not possess inversion symmetry and therefore there is no cancelation of forces exerted on the guest by the zeolite. This leads to lower diffusivity of these guest molecules.

The reason for the observed maximum in diffusivity arises from the lowered force on the guest molecule as compared to the smaller guest molecule, which encounters a higher force on itself due to the zeolite. These translate to a less undulating potential energy landscape with shallower minima and maxima for the larger guest molecule for which γ is close to unity. In the case of smaller guest molecule the larger force implies a highly undulating potential energy landscape with deep valleys and high mountains. However, in a recent report the lower force on the guest molecule at the window has not been found [29]. More studies are required to understand the origin of the observed diffusivity maximum.

The activation energy for diffusion can be obtained from an Arrhenius plot of log(D) vs. 1/T, where T is the temperature and D is the diffusivity. Variation of activation energy as a function of the guest diameter has been plotted in **Figure 6** [30]. It is seen that the activation energy is higher for the linear regime than for the anomalous regime guests. It is also seen that activation energy is maximum for the size with minimum diffusivity and minimum for the guest size in the AR with maximum diffusivity.

The observed behavior has been termed the levitation effect (LE) as its origin is in the dispersion forces, which cancel each other leading to reduced forces on the guest molecule with maximum diffusivity. Unlike diffusion of p-xylene and o- and m-xylenes whose diffusivities are controlled by the steric repulsion and therefore only p-xylene manages to diffuse, here the diffusivities are controlled by the dispersion forces, which are always attractive in nature.

Arrhenius plots for two sizes, namely, 4.96 Å and 6.0 Å are shown in **Figure 7** [28]. The smaller sized guest atom has a higher slope and activation energy than the larger sized guest atom. The activation energies for the smaller and larger sized guest atoms are respectively 5.89 kJ/mol and 3.26 kJ/mol.

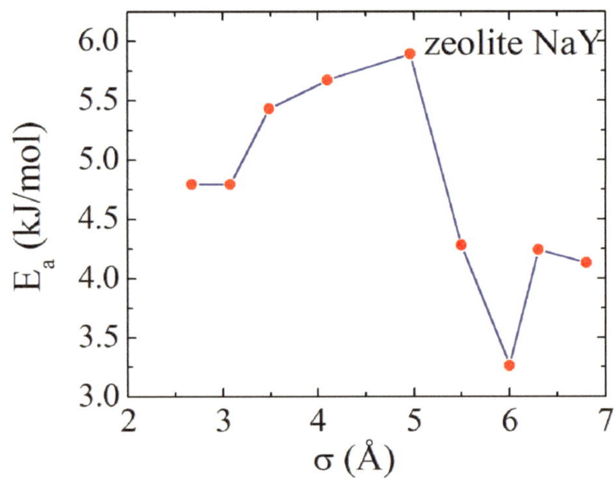

Figure 6.
Variation of activation enegy E_a as a function of vdw radius of guests, σ in zeolite NaY.

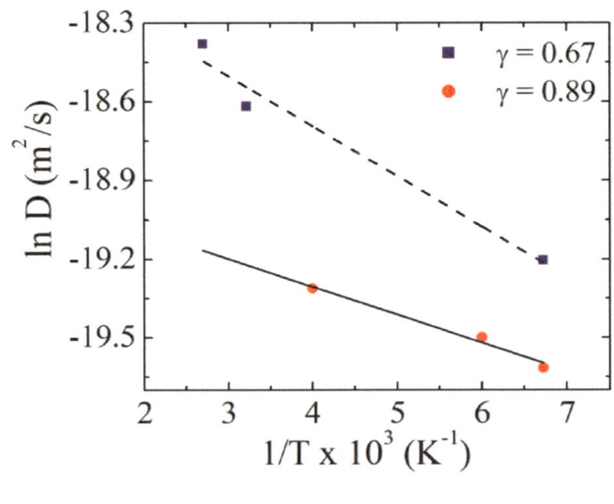

Figure 7.
Arrhenius plot of ln D for two different values of γ = 0.67 and 0.89 in zeolite NaY.

2.1 Effect of temperature on the levitation effect

The diffusivity maximum or the levitation effect is a consequence of the existence of dispersion forces. If the attractive part of the guest-zeolite interaction is switched off, then the diffusivity maximum disappears. This is shown in **Figure 8** [28]. Temperature plays an important role in the behavior of the diffusion coefficient as a function of the guest diameter.

At very higher temperatures the diffusivity maximum altogether disappears. This can be seen in **Figure 9** [31]. What determines the temperature at which the diffusivity maximum will vanish? It is the strength of interaction between the guest and the zeolite. At relatively higher temperatures when $k_BT >> U_{gz}$ the diffusivity maximum vanishes and only a monotonic dependence on guest diameter is seen.

At low temperatures, the diffusivity maximum is very pronounced with the diffusivity of the guest of 6.0 Å (in zeolite Y) showing several orders of magnitude higher value than the diffusivity of the 4.96 Å guest. This is shown in **Figure 9** [31].

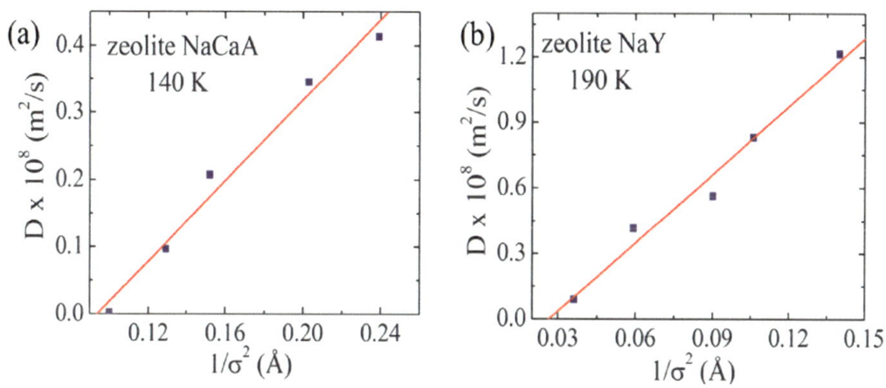

Figure 8.
Variation of diffusion coefficient D as a function of $1/\sigma^2$ in (a) zeolite NaCaA and (b) zeolite NaY, where σ is the van der Waals diameter of guest atoms. This plot is obtained without taking into account the dispersion force between guest and host atoms during molecular dynamics simulation. This clearly shows that the levitation effect is a resulting phenomenon from dispersion force of attraction.

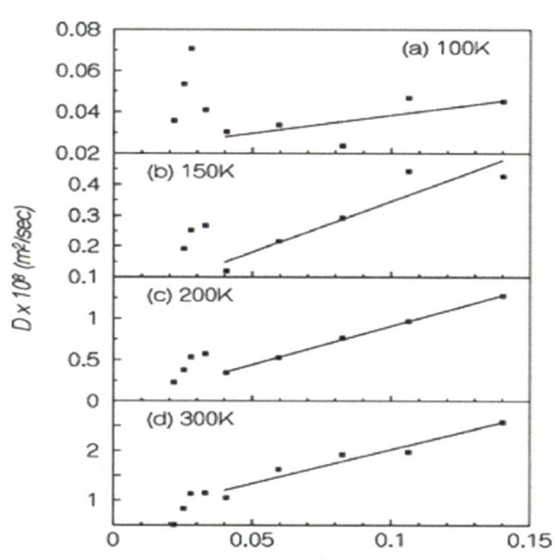

Figure 9.
Diffusion coefficient D as a function of $1/\sigma^2$ in zeolite NaY at different temperatures. It is depicted that the maxima in diffusivity disappear with increase in temperature.

At very low temperatures, the diffusivity maximum is seen to be very pronounced [32]. This can be seen from **Figure 10**. The difference between the anomalous regime guest and the linear regime guest is now several orders of magnitude. This can not be easily utilized in the practice because of the low diffusivities of both the species.

Smit and coworkers have reported a study on carbon nanotubes (CNTs). They investigated diffusion of methane in CNTs of different diameters. They found that the diffusion coefficient is maximum in the CNTs with similar diameter as the methane. They also carried out a simulation at higher temperatures when the height of the diffusivity maximum decreased and eventually disappeared similar to the disappearance of the maximum in the zeolite.

Many of these simulations have been carried out with the zeolite framework fixed. Will the diffusivity maximum persist when the framework is flexible? For this, simulations with flexible framework were carried out and the results are shown in **Figure 11** [33]. As can be seen the diffusivity maximum persists in spite of the framework flexibility. The height of the maximum is marginally lower and slightly shifted to lower values of guest diameter.

Kar and Chakravarty reported instantaneous normal mode analysis of guest of different diameters to understand the levitation effect. They could reproduce the velocity autocorrelation functions of various guest molecules in zeolite NaY [34]. Bhattacharyya and coworkers have carried out a mode coupling analysis of the levitation effect [35].

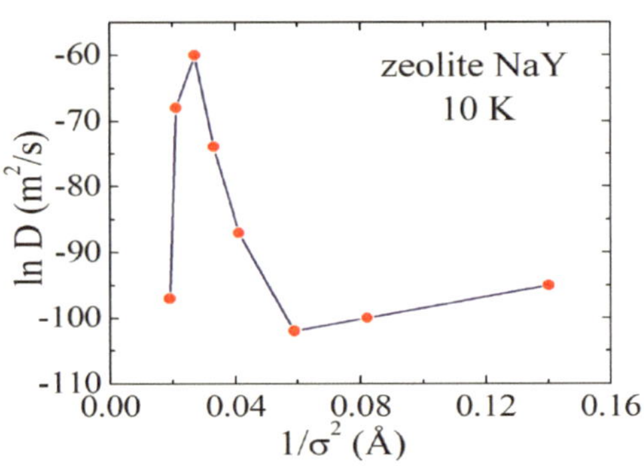

Figure 10.
Diffusion coefficient D as a function of $1/\sigma^2$ in zeolite NaY at 10 K. The diffusivity enhances to 17 orders in natural logarithmic scale.

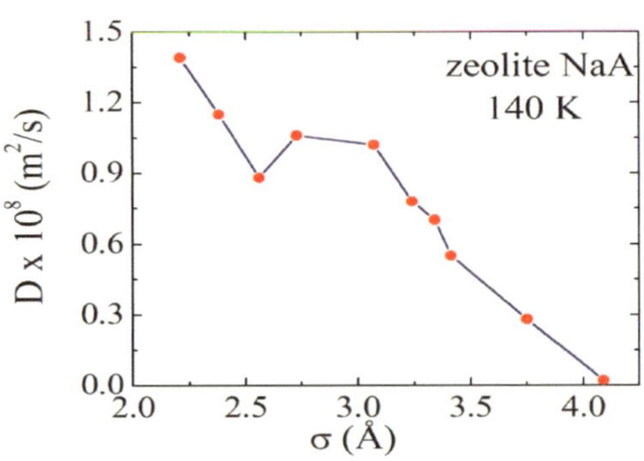

Figure 11.
Plot of diffusion coefficient D as a function of σ in flexible zeolite NaA at 140 K.

2.2 Studies on real molecular systems

Until now simulations have been on monatomic guest species diffusion in the pores of the zeolites. These guest molecules are not of interest in real laboratory or industry. Simulations were therefore carried out on hydrocarbons molecules within zeolites to see if the observed anomalous diffusion can be observed in these real hydrocarbons. Simulations were carried out on pentane isomers: n-pentane, isopentane, and neopentane. These are anisotropic molecules. Hence, the dimensions of these molecules along different directions are different. For n-pentane the direction that is relevant is the dimension of the molecule perpendicular to its long axis. The relevant dimension for the other molecule, which is isopentane, is also the dimension perpendicular to its long axis. For neopentane, which is tetrahedral in shape, the molecular diameter is the relevant dimension. There have been attempts to compute and list the various dimensions of hydrocarbon and other molecules [36]. These can be helpful in computing the γ values for various guest-zeolite systems.

Simulations of pentane isomers, n-pentane, and isopentane in AlPO-5, which has one-dimensional channels, have been reported [37]. These studies show that isopentane has a higher diffusivity as compared to n-pentane. Thus, anomalous diffusion is seen even in AlPO-5. The diffusivities obtained are $2.7 \times 10^{-8}\ m^2/s$ and $3.33 \times 10^{-8}\ m^2/s$ for n-pentane and isopentane respectively at 300 K. The potential parameters employed in this study were the unified potential parameters of Jorgensen [38]. The potential parameters are very similar to the OPLS parameters later proposed by Jorgensen. Masses of the isomers are identical and therefore the difference in the diffusivity arises from the difference in γ. For AlPO-5, it is seen that the γ values are 0.71 and 0.88 respectively for n-pentane and isopentane. The value of γ, which separates linear and anomalous regime, is around 0.75. This boundary will vary and depends on the zeolite but as a rule of thumb, a value of 0.75 may be used. The value of 0.71 lies in the linear regime while 0.88 lies in the anomalous regime. Thus, n-pentane with lower γ has a lower diffusivity, which is what one expects.

2.3 Experimental verification of the diffusivity maximum

The diffusivity of a species changes with its mass as well as other parameters such as size, temperature, etc. In simulations, the diameter of the diffusing species were changed without changing its mass. Ideally, an experimental verification of the levitation effect should do the same, that is change the diameter without changing the mass. But in real laboratory this appears almost impossible. However, Dr. S.G.T. Bhat during one of our discussions mentioned that this indeed is possible. He suggested use of isomers of a hydrocarbon all of which will have the same mass but differ in their cross-sectional diameter [39]. The choice of the experiments was also crucially important. Different techniques of measuring the diffusivity such as uptake, NMR, ZLC, or QENS yield different values of for the diffusivity of the same species. Kärger and coworkers have investigated the reasons for this [40, 41]. They have suggested that this is due to the difference in the sampling time and length scales. As MD sample over picoseconds to nanoseconds, a technique which samples for similar time scale would be ideal. As QENS samples over the same period, we choose to carry experiments with this technique. We chose zeolite NaY with three isomers of pentane, namely, n-pentane, isopentane, and neopentane. The diameters of these were calculated from their geometry and Lennard-Jones interaction parameters. Knowing the 12-ring window diameter of faujasite, we computed the levitation parameters γ for these isomers, which are 0.71, 0.86, and 0.96 for *n*-pentane, isopentane, and neopentane. The experimental QENS spectra are given in **Figure 12** [42, 43]. Also shown are the variation in half width at half maximum (HWHM) as a function of Q^2. From the broadening of the spectra as a function of Q^2, one obtains the diffusivity. The HWHM increases fastest for neopentane, followed by isopentane and last is *n*-pentane. These suggest that neopentane has the highest diffusivity, followed by isopentane and *n*-pentane in that order.

2.4 Hexane isomers in faujasite

Diffusion of hexane isomers in zeolite NaY was carried out with the help of MD technique. Linear hydrocarbons *n*-hexane (nC6), singly branched isomers 2-methyl pentane (2MP) and 3-methyl pentane (3MP) as well as the doubly branched isomers 2,2-dimethyl butane (22DMB) and 2,3-dimethyl butane (23DMB) were studied [44]. The calculated adsorption energies from MD of different isomers of hexane are listed in **Table 1**.

There is little differene in the adsorption energies of the isomers. Hence, it is difficult to separate the isomers from each other using a method based on adsorption. Instead, a kinetic-based approach might be helpful for separating the mixtures consisting of hexane isomers. In **Figure 13** the mean squared displacements of the isomers are shown for 2.25 ns. The lines are all straight suggesting good statistics [44]. The diffusivities of the isomers at these temperatures are listed in **Table 2**.

The cross-sectional diameter of the isomers can be computed from the geometry of the isomers. These are listed in **Table 3** along with the γ values for the isomers [44]. As can be seen except for *n*-hexane other isomers are all in the anomalous regime.

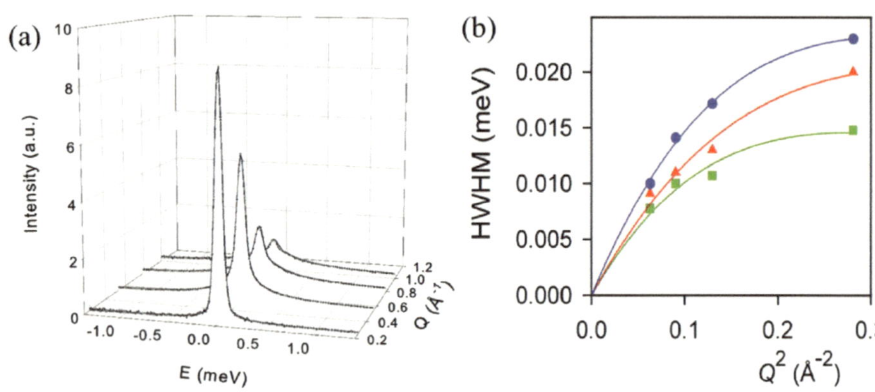

Figure 12.
(a) Spectra of QENS obtained for neopentane at 300 K for different values of the wave vector transfer Q and (b) HWHM vs. Q^2 corresponding to the translational motion of the pentane isomers in zeolite NaY at 300 K, neopentane (triangles), isopentane (squares), and n-pentane (circles).

Isomer	E_{ads} (kJ/mol)
n-Hexane	−46.7
2-Methylpentane	−45.9
3-Methylpentane	−46.4
2,3-Dimethylbutane	−45.9
2,2-Dimethylbutane	−45.8

Table 1.
Adsorption energies E_{ads} of hexane isomers in zeolite NaY.

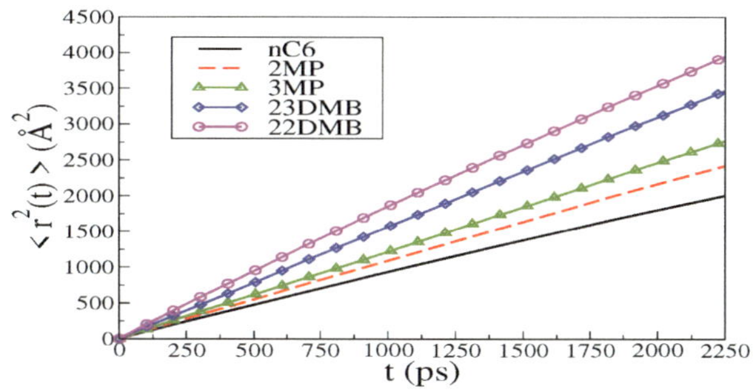

Figure 13.
Mean square displacement of hexane isomers in zeolite NaY at 250 K.

Isomer	250 K	300 K	330 K	350 K	400 K
n-Hexane	1.43(0.24)	3.59(0.35)	5.07(0.40)	7.27(0.61)	10.47(0.50)
2-Methylpentane	1.79(0.09)	3.17(0.21)	4.58(0.50)	6.44(0.55)	9.78(0.57)
3-Methylpentane	2.04(0.12)	3.92(0.24)	5.49(0.42)	6.23(0.44)	10.44(0.66)
2,3-Dimethylbutane	2.57(0.19)	4.68(0.32)	6.63(0.30)	7.26(0.61)	10.60(0.8)
2,2-Dimethylbutane	2.91(0.38)	5.67(0.29)	7.39(0.39)	7.75(0.46)	11.43(0.38)

Table 2.
Diffusion coefficients of hexane isomers in zeolite Y.

Isomer	$\sigma_\perp$ (Å)	γ
n-Hexane	6.905	0.69
2-Methylpentane	8.09	0.80
3-Methylpentane	9.55	0.95
2,3-Dimethylbutane	9.96	0.98
2,2-Dimethylbutane	10.31	1.02

Table 3.
Molecular diameter perpendicular to the long axis ($\sigma_\perp$) and the levitation parameter (γ) values of all hexane isomers, when adsorbed in zeolite NaY.

From the Arrhenius plots of diffusivity (see **Figure 14**), we have obtained the activation energies [44]. These are listed in **Table 4** [44]. It is seen that the activation energies of *n*-hexane is the highest at 11.2 kJ/mol. 2MP has an activation energy

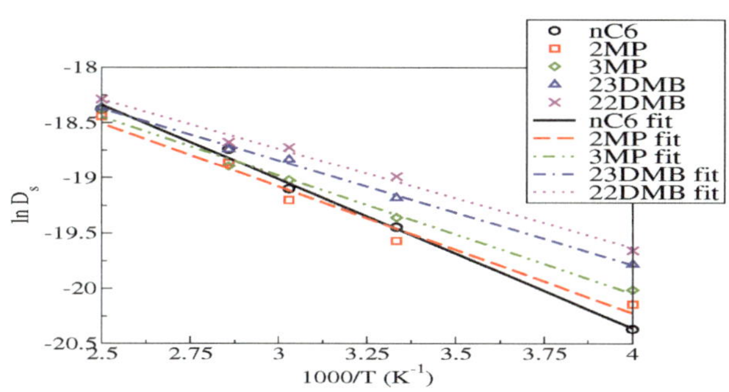

Figure 14.
Arrhenius plot of various hexane isomers in zeolite NaY.

Isomer	E_a (kJ/mol)
n-Hexane	11.2
2-Methylpentane	9.5
3-Methylpentane	8.8
2,3-Dimethylbutane	7.8
2,2-Dimethylbutane	7.4

Table 4.
Activation energies E_a of hexane isomers in zeolite NaY.

of 9.5 kJ/mol followed by 3MP (8.8 kJ/mol). The doubly branched isomers 23DMB (7.8 kJ/mol) and 22DMB (7.4 kJ/mol) have the lowest activation energy. These trends in the activation energies are according to what one expects based on the levitation effect.

The potential energy landscape of the various isomers as they diffuse during the passage through the 12-ring window of zeolite NaY is given in **Figure 15** [44]. It is seen that *n*-hexane alone has a maximum in the potential energy at the 12-ring window. 23DMB has a small maximum but overall it is a negative barrier in the vicinity of the window.

2.4.1 Kinetic-based separation of hexane isomers

Arrhenius plots of diffusivities for various isomers cross each other at some temperature. Consider two isomers. They cross at some temperature referred to as the inversion temperature. Above the temperature if one isomer has a higher diffusivity, the same isomer below the inversion temperature will have the lower diffusivity among the two isomers. The Arrhenius plots of various isomers are plotted in **Figure 16** [44]. Various pairs of isomers cross each other at different temperatures. These are listed in **Table 5** [44]. Thus, from the table it is evident that pairs 2MP/*n*-hexane have an inversion temperature of 300 K. As *n*-hexane has higher activation energy of the two, at T < 300 K, 2MP will exit from a column first and then *n*-hexane. At T > 300 K, *n*-hexane will have higher diffusivity and exit from the column first, followed by 2MP. The order of exit of the various isomers from a zeolite single crystal is seen to be different at different temperatures. There are in all 11 regions, which correspond to different order in which the isomers will exit. The order of diffusivities or the order of exit of the various isomers is listed in **Table 6** [44]. As the temperature is increased from below 300 K upto 1172 K, the order of diffusivities of various isomers is given.

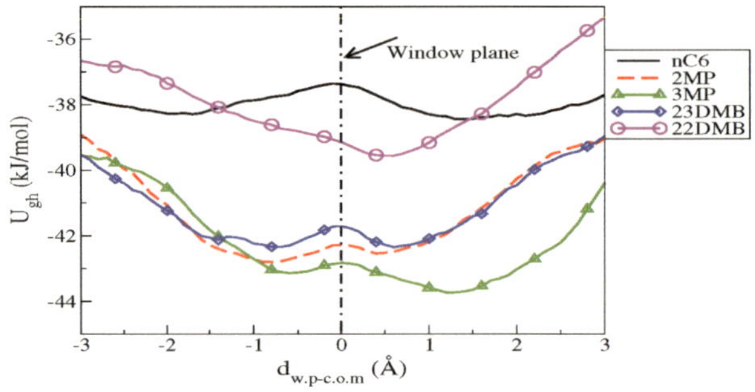

Figure 15.
Potential energy landscape for hexane isomers in zeolite NaY at 250 K. these plots are obtained by averaging over all cage-to-cage jumps.

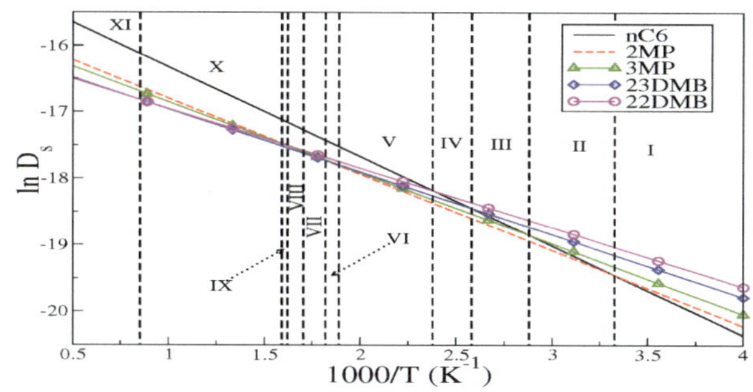

Figure 16.
Change of order of diffusivity for various hexane isomers in zeolite NaY.

Molecules inverted	T_{inv} (K)
2MP/nC6	300
MP/nC6	347
DMB/nC6	387
DMB/nC6	418
DMB/3MP	528
DMB/2MP	549
MP/2MP	586
DMB/2MP	613
DMB/3MP	627
DMB/23DMB	1172

Table 5.
Inversion temperatures for diffusivities (T_{inv}) of hexane isomers inside zeolite NaY.

Region	Temp range (K)	Order of diffusivities
I	T < 300	nC6 < 2MP < 3MP < 23DMB < 22DMB
II	300,347	2MP < nC6 < 3MP < 23DMB < 22DMB
III	347,387	2MP < 3MP < nC6 < 23DMB < 22DMB
IV	387,418	2MP < 3MP < 23DMB < nC6 < 22DMB
V	418,528	2MP < 3MP < 23DMB < 22DMB < nC6
VI	528,549	2MP < 23DMB < 3MP < 22DMB < nC6
VII	549,586	23DMB < 2MP < 3MP < 22DMB < nC6
VIII	586,613	23DMB < 3MP < 2MP < 22DMB < nC6
IX	613,627	23DMB < 3MP < 22DMB < 2MP < nC6
X	6,271,172	23DMB < 22DMB < 3MP < 2MP < nC6
XI	T > 1172	22DMB < 23DMB < 3MP < 2MP < nC6

Table 6.
Different regions according to order of diffusivities of hexane isomers in zeolite NaY.

2.5 End-to-end length of *n*-hexane during diffusion

An interesting observation from the study was the variation of end-to-end length, $L_{e\text{-}e}$ of *n*-hexane during diffusion. The $L_{e\text{-}e}$ was computed at various temperatures for *n*-hexane. It showed that there is a decrease in the end-to-end length of *n*-hexane as the temperature increases. A plot of average $L_{e\text{-}e}$ as a function of the temperature is shown in **Figure 17** [44]. The distribution of the $L_{e\text{-}e}$ has also been computed for various temperatures and these are shown in **Figure 18** [44]. As can be seen at higher temperatures the distribution has a higher probability for smaller values of $L_{e\text{-}e}$, between 3 and 5 Å and less probability for higher values (5–6.4 Å) of $L_{e\text{-}e}$. This is what leads to a decrease in average $L_{e\text{-}e}$ with temperature. These suggest that *n*-hexane curls up at higher temperatures and has a higher population of gauche conformers. In order to check if this was true, we have obtained the number of gauche conformer population and these are listed in **Table 7**.

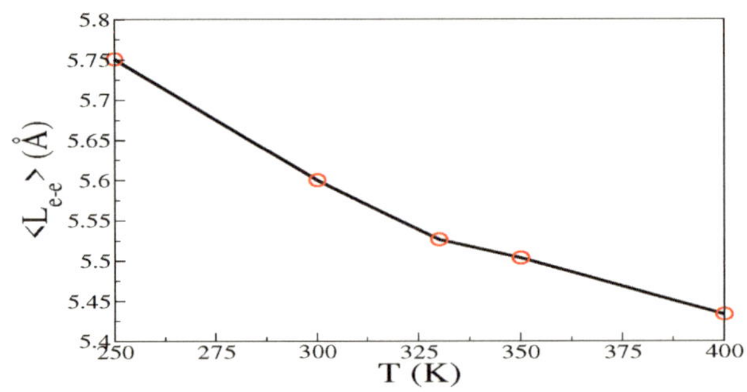

Figure 17.
Variation of end-to-end length of n-hexane in zeolite NaY as a function of temperature.

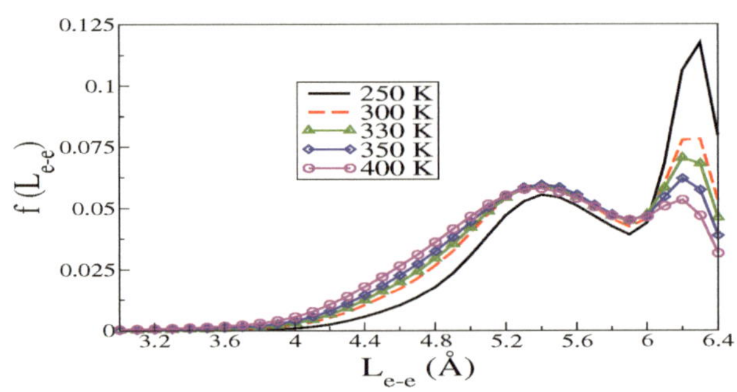

Figure 18.
Distribution of end-to-end length n-hexane in zeolite NaYat various temperatures.

Temp (K)	ϕ_1	ϕ_2	ϕ_3
250	24.9	19.0	24.7
300	29.5	23.8	29.4
330	32.3	26.4	32.5
350	34.3	28.2	34.3
400	36.5	30.4	36.6

Table 7.
% gauche conformations for each dihedral angle of nC6 inside zeolite NaY at various temperatures.

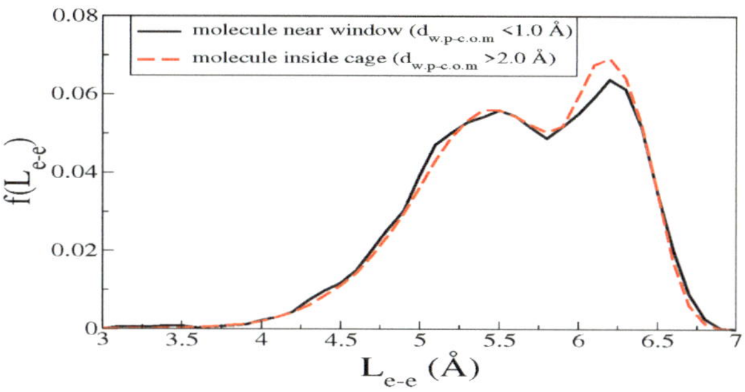

Figure 19.
Variation of the distribution of end-to-end length of nC6 in zeolite NaY.

The distribution of end-to-end length of *n*-hexane is shown in **Figure 19** for two specific situations: (i) when the center of mass of *n*-hexane is close to the 12-ring window (within ±1.0 Å) and (ii) when the center of mass is greater than 2 Å from the window plane [44]. As can be seen, the probability for smaller values of $L_{e\text{-}e}$ is higher when the molecule of *n*-hexane is closer to the window. Thus, it appears that the *n*-hexane curls up trying to increase its cross-sectional diameter while passing through the 12-ring window. This leads to a slightly lower value of energetic barrier for *n*-hexane when it curls up.

2.6 Separation using levitation effect

Using the levitation effect, separation of molecular mixtures can be realized. This approach differs from the usual approach toward separation. In the usual approach separation is achieved because the smaller sized molecules generally diffuse faster as compared to larger sized guest molecules. Or alternately, certain molecules enter the pores while bigger sized molecules do not even enter the pore network. In this case only those which enter the pores are able to pass through the column of the zeolite while others are unable to pass through leading to good separation.

Consider now a binary mixtures in which both the components are able to enter the pore network. When we use the levitation effect, both mixtures will diffuse but here the larger sized guest will diffuse faster provided the bottleneck of the zeolite has a dimaeter that is comparable to the diameter of the larger sized guest molecule. The smaller diameter guest molecule will diffuse slower and this leads to separation of the components.

In all these separations, however, one thing that is common is that both the components (in the case of a binary mixture) will diffuse in the same direction. This leads at best to a reasonable degree of separation. However, much higher degree of separation can be achieved if the two components move in opposite directions. We have devised a novel approach to separation in which the two components move in opposite directions [45]. In this approach, the fact that the guest in AR and the LR regimes have out-of-phase potential energy landscape is utilized. This is shown in **Figure 20** [45]. In addition, a hot zone is placed to the left of the 12-ring window in the system consisting of zeolite A with argon (of AR) and neon (of LR). The hot zone drives argon toward the left and the neon toward to right. The result is a very high degree of separation. This has been demonstrated through nonequilibrium Monte Carlo simulations with inhomogeneous temperature [45]. In **Figure 21** a plot of the separation factor as a function of Monte Carlo steps is shown. Normally separation factors that are achieved are of the order of 2000–4000. But here we see that separation factors of the order of $>10^{8-10}$ can be realized.

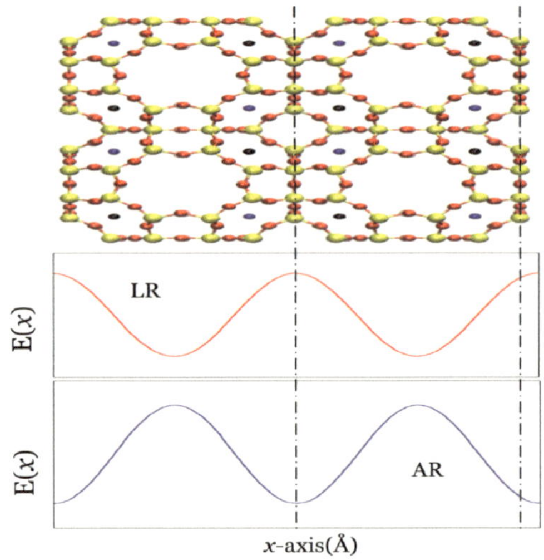

Figure 20.
Schematic of potential energy landscapes of AR (BLUE) and LR (RED) particles are shown below the unit cell crystal structure of zeolite NaCaA. The black dotted lines are the window planes of the zeolite.

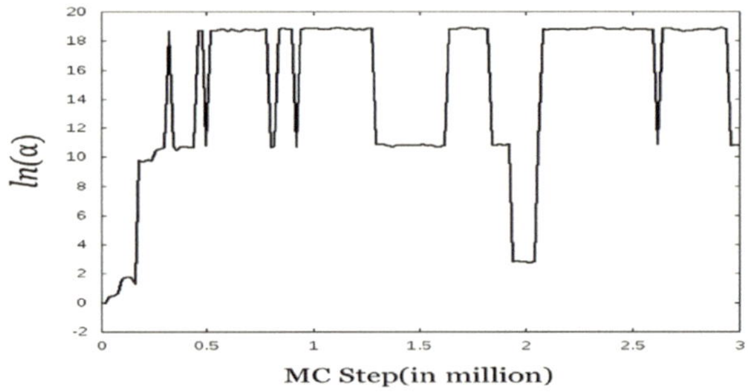

Figure 21.
Evolution of separation factor with Monte Carlo steps. This shows that AR particles can be fully separated from LR particles using this separation technique.

3. Conclusions

The surprising anomaly in the diffusivity of a guest diffusing within the confined regions of a zeolite or a CNT or any other porous solid has many uses. It also explains the observed increase in ionic conductivity as a function of ionic radius on going from Li^+ to Na^+ to K^+ and Rb^+ and Cs^+ [46]. Levitation effect was first observed in the experiments by Kemball and later Derouane showed its existence through theory [47–49]. It can be used in the separation of molecular mixtures as well as in separations of very high degree. Levitation effect can be better put into practice with single crystals and nano devices.

Acknowledgements

The authors wish to acknowledge support from Department of Science and Technology, New Delhi and computational facilities from Thematic Unit of Excellence for Computational Materials Science established with support from Nano Mission, DST, New Delhi.

DOI: http://dx.doi.org/10.5772/intechopen.92685

Author details

Shubhadeep Nag and Yashonath Subramanian*
Solid State and Structural Chemistry Unit, Indian Institute of Science, Bengaluru, India

*Address all correspondence to: yashonath@iisc.ac.in

References

[1] Boronat M, Corma A. What is measured when measuring acidity in zeolites with probe molecules? ACS Catalysis. 2019;**9**(2):1539-1548

[2] Moliner M, Corma A. From metal-supported oxides to well-defined metal site zeolites: The next generation of passive nox adsorbers for low-temperature control of emissions from diesel engines. Reaction Chemistry & Engineering. 2019;**4**:223-234

[3] Li C, Moliner M, Corma A. Building zeolites from precrystallized units: Nanoscale architecture. Angewandte Chemie International Edition. 2018; **57**(47):15330-15353

[4] Bereciartua PJ, Cantín Á, Corma A, Jordá JL, Palomino M, Rey F, et al. Control of zeolite framework flexibility and pore topology for separation of ethane and ethylene. Science. 2017; **358**(6366):1068-1071

[5] Sastre G, Kärger J, Ruthven DM. Diffusion path reversibility confirms symmetry of surface barriers. The Journal of Physical Chemistry C. 2019; **123**(32):19596-19601

[6] Toda J, Sastre G. Diffusion of trimethylbenzenes, toluene, and xylenes in uwy zeolite as a catalyst for transalkylation of trimethylbenzenes with toluene. The Journal of Physical Chemistry C. 2018;**122**(14):7885-7897

[7] Zachariou A, Hawkins A, Lennon D, Parker SF, Suwardiyanto SK, Matam CRA, et al. Investigation of zsm-5 catalysts for dimethylether conversion using inelastic neutron scattering. Applied Catalysis A: General. 2019;**569**:1-7

[8] Caro J, Kärger J. From computer design to gas separation. Nature Materials

[9] Kumar P, Kim DW, Rangnekar N, Xu H, Fetisov EO, Ghosh S, et al. One-dimensional intergrowths in two-dimensional zeolite nanosheets and their effect on ultra-selective transport. Nature Materials. 2020;**19**:443-449

[10] Quesne MG, Silveri F, de Leeuw NH, Catlow CRA. Advances in sustainable catalysis: A computational perspective. Frontiers in Chemistry. 2019;**7**:182

[11] Kang JH, Alshafei FH, Zones SI, Davis ME. Cage-defining ring: A molecular sieve structural indicator for light olefin product distribution from the methanol-to-olefins reaction. ACS Catalysis. 2019;**9**(7):6012-6019

[12] Dusselier M, Davis ME. Small-pore zeolites: Synthesis and catalysis. Chemical Reviews. 2018;**118**(11):5265-5329

[13] Davis ME. A thirty-year journey to the creation of the first enantiomerically enriched molecular sieve. ACS Catalysis. 2018;**8**(11):10082-10088

[14] Barrer RM. Hydrothermal Chemistry of Zeolites. London: Academic Press; 1982

[15] Kärger J, Ruthven DM. Diffusion in Zeolites and Other Microporous Solids. New York, USA: John Wiley and Sons Inc.; 1992

[16] Dong J, Lin YS, Liu W. Multicomponent hydrogen/hydrocarbon separation by mfi-type zeolite membranes. AIChE Journal. 2000;**46**(10):1957-1966

[17] Arruebo M, Falconer JL, Noble RD. Separation of binary c5 and c6 hydrocarbon mixtures through mfi zeolite membranes. Journal of Membrane Science. 2006;**269**(1):171-176

[18] Seader JD, Henley EJ. Separation Process Principles. 1st ed. New York, USA: John Wiley and Sons Inc.; 1998

[19] Colella C. Ion exchange equilibria in zeolite minerals. Mineralium Deposita. 1996;**31**(6):554-562

[20] Loiola A, Andrade J, Sasaki J, da Silva L. Structural analysis of zeolite naa synthesized by a cost-effective hydrothermal method using kaolin and its use as water softener. Journal of Colloid and Interface Science. 2012; **367**(1):34-39

[21] Smit B, Maesen TLM. Molecular simulations of zeolites: Adsorption, diffusion, and shape selectivity. Chemical Reviews. 2008;**108**(10): 4125-4184

[22] Tarditi AM, Horowitz GI, Lombardo EA. Xylene isomerization in a zsm-5/ss membrane reactor. Catalysis Letters. 2008;**123**(1):7-15

[23] Young L, Butter S, Kaeding W. Shape selective reactions with zeolite catalysts: Iii. Selectivity in xylene isomerization, toluene-methanol alkylation, and toluene disproportionation over zsm-5 zeolite catalysts. Journal of Catalysis. 1982;**76**(2): 418-432

[24] Dubbeldam D, Calero S, Maesen TLM, Smit B. Understanding the window effect in zeolite catalysis. Angewandte Chemie International Edition. 2003;**42**(31):3624-3626

[25] Kärger J, Petzold M, Pfeifer H, Ernst S, Weitkamp J. Single-file diffusion and reaction in zeolites. Journal of Catalysis. 1992;**136**(2): 283-299

[26] Yashonath S, Ghorai PK. Diffusion in nanoporous phases: Size dependence and levitation effect. The Journal of Physical Chemistry B. 2008;**112**(3): 665-686

[27] Yashonath S, Santikary P. Influence of non-geometrical factors on intracrystalline diffusion. Molecular Physics. 1993;**78**(1):1-6

[28] Yashonath S, Santikary P. Diffusion of sorbates in zeolites y and a: Novel dependence on sorbate size and strength of sorbate-zeolite interaction. The Journal of Physical Chemistry. 1994; **98**(25):6368-6376

[29] Sharma A, Ghorai PK. Effect of force and location of bottleneck for particle moving through window under encapsulation. Journal of Chemical Science. 2017;**129**(8):1293-1300

[30] Rajappa C, Yashonath S. Levitation effect and its relationship with the underlying potential energy landscape. The Journal of Chemical Physics. 1999; **110**(12):5960-5968

[31] Yashonath S, Rajappa C. Temperature dependence of the levitation effect implications for separation of multicomponent mixtures. Faraday Discussions. 1997;**106**:105-118

[32] Ghorai PK, Yashonath S. Diffusion anomaly at low temperatures in confined systems from the rare events method. The Journal of Physical Chemistry B. 2004;**108**(22):7098-7101

[33] Santikary P, Yashonath S. Dynamics of zeolite cage and its effect on the diffusion properties of sorbate: Persistence of diffusion anomaly in naa zeolite. The Journal of Physical Chemistry. 1994;**98**(37):9252-9259

[34] Kar S, Chakravarty C. Instantaneous normal mode analysis of the levitation effect in zeolites. The Journal of Physical Chemistry B. 2000;**104**(4):709-715

[35] Nandi MK, Banerjee A, Bhattacharyya SM. Non-monotonic size dependence of diffusion and levitation effect: A mode-coupling theory analysis. The Journal of Chemical Physics. 2013; **138**(12):124505

[36] Federico J-C, Laredo GC. Molecular size evaluation of linear and branched paraffins from the gasoline pool by dft

quantum chemical calculations. Fuel. 2004;**83**:2183-2188

[37] Bhide SY, Kumar AVA, Yashonath S. Diffusion of hydrocarbons in confined media: Translational and rotational motion. Journal of Chemical Sciences. 2001;**113**(5):559-577

[38] Jorgensen WL, Madura JD, Swenson CJ. Optimized intermolecular potential functions for liquid hydrocarbons. Journal of the American Chemical Society. 1984;**106**(22): 6638-6646

[39] Personal communatication with Dr. S.G. Thirumaleshwara Bhat.

[40] Kärger J, Ruthven DM. On the comparison between macroscopic and n.m.r. measurements of intracrystalline diffusion in zeolites. Zeolites. 1989;**9**(4): 267-281

[41] Kärger J, Ruthven DM. Diffusion in nanoporous materials: Fundamental principles, insights and challenges. New Journal of Chemistry. 2016;**40**: 4027-4048

[42] Borah BJ, Jobic H, Yashonath S. Levitation effect in zeolites: Quasielastic neutron scattering and molecular dynamics study of pentane isomers in zeolite nay. The Journal of Chemical Physics. 2010;**132**(14):144507

[43] Jobic H, Borah BJ, Yashonath S. Neutron scattering and molecular dynamics evidence for levitation effect in nanopores. The Journal of Physical Chemistry B. 2009;**113**(38):12635-12638

[44] Thomas AM, Subramanian Y. Hexane isomers in faujasite: Anomalous diffusion and kinetic separation. The Journal of Physical Chemistry C. 2017; **121**(27):14745-14756

[45] Anil Kumar AV, Yashonath S, Ananthakrishna G. Separation of mixtures at nano length scales: Blow torch and levitation effect. The Journal of Physical Chemistry B. 2006;**110**(8): 3835-3840

[46] Ghorai PK, Yashonath S. Evidence in support of levitation effect as the reason for size dependence of ionic conductivity in water: A molecular dynamics simulation. The Journal of Physical Chemistry B. 2006;**110**(24): 12179-12190

[47] Kemball C. Entropy of adsorption. Advances in Catalysis. 1950;**2**:233-250

[48] Derouane EG. The energetics of sorption by molecular sieves: Surface curvature effects. Chemical Physics Letters. 1987;**142**(3):200-204

[49] Derouane EG, Andre J-M, Lucas AA. Surface curvature effects in physisorption and catalysis by microporous solids and molecular sieves. Journal of Catalysis. 1988;**110**(1): 58-73

Section 2

Layered Zeolites

Chapter 5

New Trends in Layered Zeolites

Hao Xu and Peng Wu

Abstract

Layered zeolites, with a flexible and changeable interlayer connection, can be modified to give a great number of derivative structures with enlarged pore sizes or enhanced external surface areas, via swelling, delamination, pillaring, or silylation. In recent years, great efforts have been devoted to the synthesis of novel-layered zeolite precursors, by using the specially designed bifunctional amphiphilic surfactants as the structure-directing agents or through the selective degradation of double four ring-containing germanosilicates. In addition, the novel modifications, such as mild delamination, interlayer expansion assisted by deconstruction-reconstruction, and layer-stacking reorganization by dissolution-recrystallization, have also been developed to create more derivatives while achieving better preservation of layer structures. Recent progresses in the field of layered zeolites are summarized in this chapter, and the challenges for future development are also proposed.

Keywords: layered zeolite, design synthesis, structural modification, delamination, interlayer expansion

1. Introduction

Zeolites, with high crystalline three-dimensional (3D) frameworks composed of TO_4 (T = Si or Al, etc.) tetrahedrons, have found their unique advantages in the field of adsorption, separation, and catalysis, owing to a large surface area, uniform pore channels of molecular size, strong acidity, and redox ability. According to the International Zeolite Association (IZA), 244 kinds of zeolites have been recognized up to date, most of which are hydrothermally synthesized [1]. In addition, the topotactic conversion from two-dimensional (2D) lamellar precursors to 3D rigid zeolite framework contributes ~10% of the whole zeolite family [2]. Rather than strong and rigid covalent bonds, relative weak and flexible hydrogen bonds are the interaction force between the neighboring layers in 2D lamellar precursors, which are also called layered zeolites. These hydrogen bonds are derived from the abundant silanol groups on the layer surface, which would condense to form Si—O—Si linkage upon calcination and then produce 3D zeolite frameworks.

In the very beginning, layered zeolites are occasionally obtained from the traditional synthetic gels that were designed to produce 3D zeolites. However, the formation mechanism of these layered zeolites is still a mystery, which prohibits the researchers to design and synthesize more novel-layered zeolites via the traditional hydrothermal synthesis. Recently, several novel strategies have been proposed to synthesize layered zeolite, including the usage of specially designed bifunctional structure-directing agents (SDA) [3] and the transformation of 3D germanosilicates to 2D lamellar zeolites by posttreatment [4]. The newly established methods expanded the layered zeolite family, and there are now nearly 30 kinds of layered zeolites available (**Table 1**).

IntechOpen

The interlayer flexible hydrogen bonds endowed the layered zeolites with modifiable structural property. Post-modifications including swelling [29], delamination [30, 31], pillaring [32], silylation [33], and detemplating [34] have been reported to increase the interlayer space or to gain higher external surface area (**Figure 1**). The classical swelling process is achieved by the intercalation of layered zeolites with large-sized surfactant molecules in the alkaline organic ammonium solution, resulting in enlarged interlayer space [29]. An ultrasonic treatment over the swollen intermediate produces full delaminated materials, changing the original ordered stacking style to house-of-cards arrangement and greatly enhancing the external surface area [30, 31]. Again, based on the swollen intermediate, the enhanced interlayer space can be stabilized by rigid silica pillars, giving pillaring materials with interlayer mesopores and intralayer micropores [32]. In contrast, interlayer silylation is an atom-level accurate modification to produce interlayer-expanded structures [33]. Detemplating, including the full and partial removal of the interlayer SDA molecules, results in a 3D zeolite framework and partial delaminated

Layered precursors	Structure code[a]	Main pores for 3D structure	Reference
MCM-22	MWW	10 × 10-R, 12R cages	[5]
PSH-3			[6]
SSZ-25			[7]
ITQ-1			[8]
ERB-1			[9]
EMM-10			[10]
SSZ-70			[11]
ECNU-5			[12]
PREFER	FER	10 × 8-R	[13]
PLS-3			[14]
MCM-47	CDO	8 × 8-R	[15]
MCM-65			[16]
PLS-1			[17]
PLS-4			[14]
RUB-36			[18]
EU-19	CAS	8R	[19]
Nu-6(1)	NSI	8 × 8-R	[20]
RUB-15	SOD	6R	[21]
RUB-18	RWR	8 × 8-R	[22]
RUB-39	RRO	10 × 8-R	[23]
PreAFO	AFO	10R	[24]
Lamellar MFI	MFI	10 × 10-R	[3]
IPC-1P	PCR	10 × 8-R	[4]
	*PCS	12 × 10-R, 10 × 8-R	[25]
ECNU-21P	—	10 × 6-R	[26]
MCM-69(P)	—	—	[27]
HUS-2	—	—	[28]

[a] *The structure code for the 3D zeolite obtained from layered precursor upon calcination.*

Table 1.
Listing of existing layered zeolite.

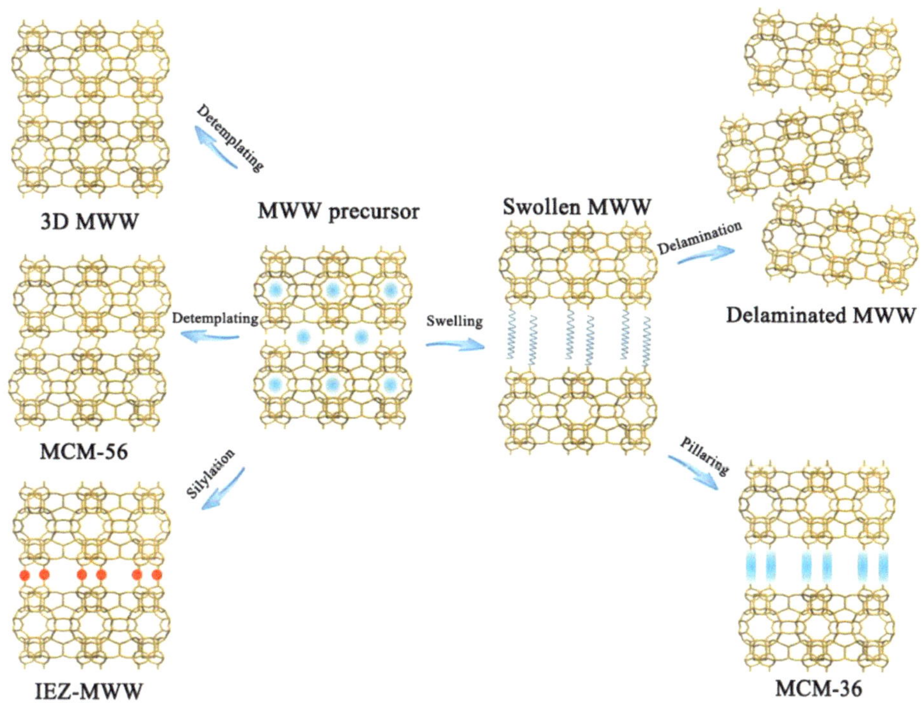

Figure 1.
Summary of the post-synthesis modifications over a typical MWW lamellar zeolite.

materials, respectively [34]. Although these posttreatments have achieved a great success for most of the layered zeolites, the corrosion effect of the alkaline solution on intralayer structure in swelling process, the multistep procedure of delamination, and the failure of silylation over some special layered zeolites cannot be ignored, and new modification strategies are highly desired.

This chapter will review the recently developed synthetic methods, including the usage of bifunctional surfactants and the selective degradation of germanosilicates, and the modification strategies of layered zeolites, such as mild delamination, interlayer expansion assisted by deconstruction-reconstruction, and layer-stacking reorganization by dissolution-recrystallization. One of the most important driving forces of novel-layered zeolite exploration and modification improvement is to design synthesis of more efficient catalysts, with better accessibility of active sites provided by enlarged pores and external surface areas. Thus, this chapter will also cover the catalytic performance of the newly developed layered zeolite and derivative materials in the solid acid^{-}/base-catalyzed reactions and the liquid selective oxidation reactions. For the conventional synthesis, characterization, classical modifications, and applications of layered zeolites, the readers could find them in the published reviews [35–40]. In the end, the challenge and possible future development in the field of layered zeolites is prospected.

2. Newly developed synthetic methods

2.1 Synthesis of layered zeolites with bifunctional amphiphilic SDAs

Compared to the continuous expanding in the 3D directions for typical zeolite frameworks, the layered structures only spread in 2D directions, with the growth in the third direction being interrupted. Considering this unique structural property of layered zeolites, Ryoo et al. proposed a novel strategy to synthesize layered zeolites using specially designed bifunctional amphiphilic SDAs with the hydrophilic diquaternary ammonium head to direct the crystallization of intralayer structures and on the other hand the hydrophobic long alkyl chain to prevent the continuous growth in the direction vertical to the layers [3, 41]. MFI-layered zeolite, hardly obtained in the traditional hydrothermal synthesis, was firstly reported using this method with the surfactant of $C_{22}H_{45}—N^+(CH_3)_2—C_6H_{12}—N^+(CH_3)_2—C_6H_{13}$. A large amount of Na^+ in the synthetic gel favored the alternating stacking of 2-nm-thick MFI nanosheets and 2.8-nm-thick surfactant micelles, producing multilamellar MFI with an overall thickness of 30–40 nm (**Figure 2A–D**). However, reducing the Na^+ resulted in the formation of unilamellar MFI nanosheets (**Figure 2E**). Both of the two-layered MFI zeolites exhibited a significantly higher surface area than that of the traditional 3D MFI zeolite, due to the formation of mesopores upon calcination. The layered MFI zeolites showed longer lifetime in the methanol-to-gasoline reaction and higher catalytic activity in the reactions involving large-size molecules, due to the simultaneous presence of mesopores and micropores. Similar superior catalytic results were reported by comparing the performance of layered TS-1 zeolite to that of bulk TS-1 in the epoxidation reactions [42]. The as-synthesized multilamellar MFI zeolite has also been applied as an acid-base bifunctional catalyst in the Knoevenagel condensation reactions. The acid site was derived from the Al-related Brönsted acidity, while the ammonium group located in the pore mouth served as a base site [43].

The structure of the bifunctional amphiphilic SDAs was then proved to be critical for the synthesis of layered MFI zeolite [44]. Too small space between the two ammonium groups would result in bulk MFI, while too large space leads to the disordered stacking of MFI nanosheets. Tuning the number of ammonium group is effective in controlling the thickness of MFI nanosheets. The hydrophobic alkyl chain should be long enough to form a micellar structure. Once two long alkyl chains were present in the bifunctional SDA, hexagonal mesostructure was formed with the pore wall characterized with crystalline MFI structure having the thickness of 1.7 nm [45]. By introducing aromatic groups into the hydrophobic chain of bifunctional SDA molecules, Che et al. have synthesized several lamellar MFI zeolites with different mesostructures, benefiting from the strong π-π stacking and the geometrical match between aromatic groups and MFI zeolitic frameworks [46–48]. This strategy of using bifunctional SDAs has been successfully extended to

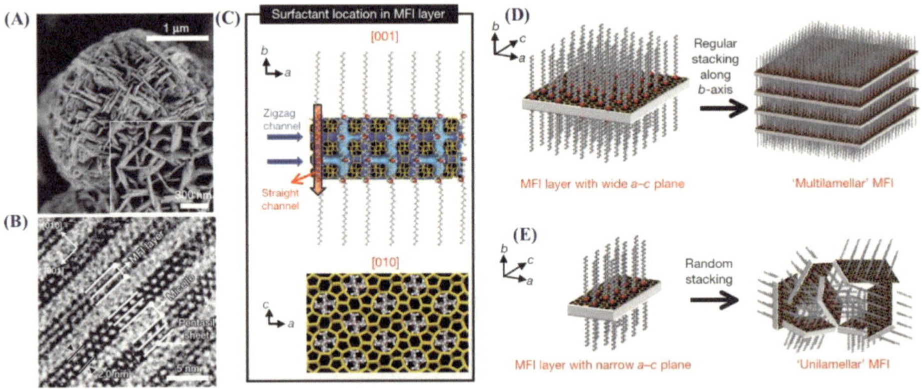

Figure 2.
SEM (A) and TEM (B) images of layered MFI zeolite. The scheme for the single layer MFI nanosheet (C) and the structural model for multilamellar MFI (D) and unilamellar MFI (E).

the synthesis of nanosheets with MTW [3], *MRE [49], and AlPO [50] frameworks, which have never been synthesized as 2D zeolites before.

For the well-known MWW-layered zeolite, a bifunctional SDA with the head group resembling the SDA for conventional MWW-layered zeolite and a long alkyl chain connected by diquaternary ammonium linker directed the crystallization of MIT-1 composed of MWW nanosheets with a house-of-cards arrangement (**Figure 3**), which was similar to the delaminated MWW material [51]. MIT-1 exhibited high mesoporosity with the external surface area higher than that of MCM-56 but lower than ITQ-2. This one-pot synthetic method using rational designed bifunctional SDA to prepare delaminated material avoids the conventional multistep process and also the amorphization of layer structure in the swelling treatment. Another directly synthesized delaminated MWW zeolite, DS-ITQ-2, was reported by Corma et al., under the co-directing effect of hexamethyleneimine (HMI) and a bifunctional SDA of *N*-hexadecyl-*N*′-methyl-DABCO ($C_{16}DC_1$) [52]. HMI, the traditional template for the crystallization of layered MWW zeolite, was applied to direct the crystallization of MWW layers. For the bifunctional SDA, the hydrophilic head is located in the pocket of MWW layers, while the long hydrophobic chain prevents the ordering of stacking along *c* axis. The obtained DS-ITQ-2 showed a comparable catalytic activity as the conventional delaminated material of ITQ-2 in the liquid phase alkylation of benzene with propene.

2.2 Synthesis of layered zeolites by selective removal of double four rings from germanosilicates

Germanosilicates, with novel topologies and large-pore channels, have shown their great potential in the catalytic reactions involving bulky substrates. Ge atoms favored the formation of double four ring (D4R) and double three ring (D3R) when building the zeolite frameworks together with silica atoms, due to their longer

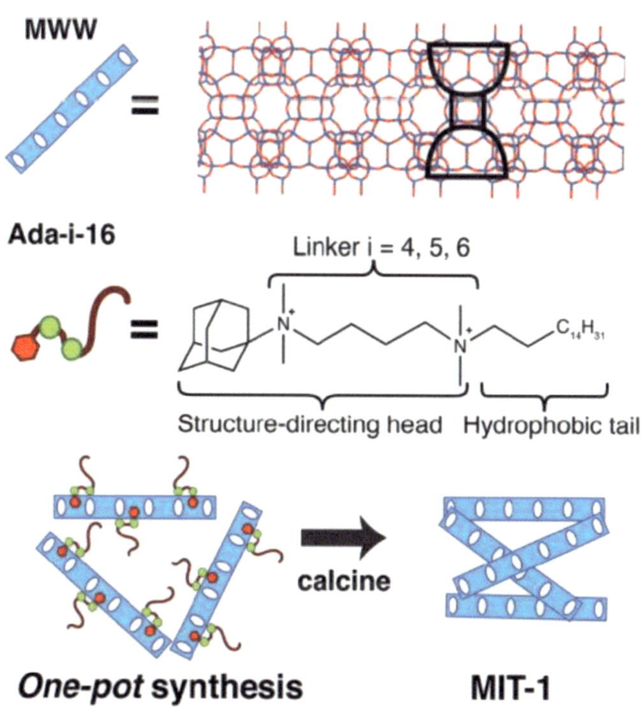

Figure 3.
Scheme description of one-pot synthesis of MIT-1 with delaminated MWW structure.

Ge—O bond and smaller Ge—O—Ge bond angle. Then, it is easy for these D4R and D3R subunits to construct large or extra-large pore channels, such as ITQ-37 (30R) [53], ITQ-43 (28R) [54], ITQ-44 (16R) [55], etc. However, the instability of Si(Ge)—O—Ge bonds prevents the practical applications of germanosilicates. On the bright side, the instability of germanosilicates endowed them structural modifiable properties. Taking full advantage of this, Čejka et al. have put forward an effective strategy, called the assembly-disassembly-organization-reassembly (ADOR), to transform the 3D zeolite frameworks to 2D layered zeolite by selectively removing the Ge-rich D4R subunits in UTL germanosilicate under acidic environment to give IPC-1P lamellar intermediate (**Figure 4**) [4]. The obtained lamellar zeolite was then organized and reassembled by organic amine treatment, calcination, or interlayer silylation, giving a series of novel zeolite frameworks (IPC-*n*) [56–58]. Three of these novel UTL-derived zeolites, including OKO [59], PCR [4], and *PCS [25], have been recognized by the IZA structure committee. The alternating stable Si-rich zeolitic layer and instable Ge-rich D4R subunits and high Ge content are the key

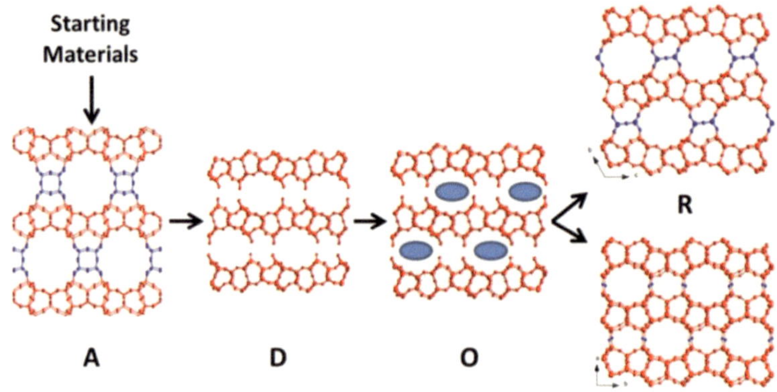

Figure 4.
Scheme description of ADOR strategy to create layered zeolite intermediates under acidic condition and novel germanosilicate derivatives.

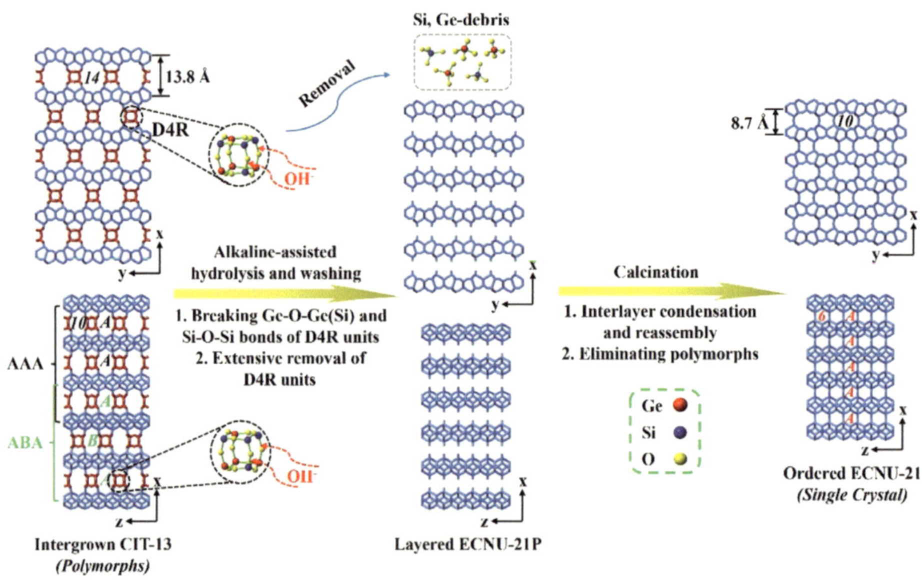

Figure 5.
Selective hydrolysis of germanosilicates CIT-13 under alkaline medium to give layered zeolite ECNU-21P.

factors for the successful structural transformation. The ADOR strategy assisted by an acidic solution has been successfully extended to other germanosilicates, including ITH, IWR [60], IWW [61], UOV [62], and SAZ-1 [63].

Although CIT-13 germanosilicate meets all the criteria for ADOR strategy, it failed to give any lamellar zeolite. More detailed characterization indicated that the presence of Si—O—Si bonds vertical to the Si-rich layers may prohibit the successful selective removal of D4R subunits [64]. Wu et al. reported a mild alkaline treatment to hydrolyze both Ge—O—Ge(Si) and Si—O—Si bonds in the D4R subunits, giving a layered intermediate ECNU-21P (**Figure 5**), which was transformed to 3D ECNU-21 zeolite upon calcination [26]. In addition, the selective removal of D4R subunits from CIT-13 zeolite eliminates the intergrowth phenomena and resulted in a single crystalline zeolite framework. The alkaline medium-assisted ADOR strategy works as a good addition to the original acid system and is expected to create more layered zeolites.

3. Novel modification strategies

3.1 Delamination of layered zeolite under mild conditions

The conventional delamination process includes the swelling process and a subsequent ultrasound treatment in alkaline solution with the typical pH value of ~13, which causes severe dissolution of the intralayer structures [30, 31]. Moreover, the mesoporous phase formed form dissolved silica species in swelling process, like MCM-41, may also contribute to the high surface area, which could give cheating information for the success delamination. Zones and Katz reported a delaminated MWW zeolite synthesized under mild condition, where a solution of tetrabutylammonium fluoride and chloride surfactants with a pH value of 9 was used (**Figure 6**) [65]. The fluoride ion could form a strong interaction with Si atoms, while the chloride ions would attack Al atoms in the delamination process. High solid yield of ~90% was obtained under the mild condition compared to the yield of ~75% under classical alkaline condition with the same treatment temperature and duration. The successful delamination was strongly evidenced by the loss of interlayer 10R pores, characterized by the decrease of the N_2 uptake in the relative pressure of $10^{-7} < p/p_0 < 10^{-4}$. The absence of Q^2 signals and the sharper Q^4 resonance bands in the ^{29}Si NMR spectrum of delaminated UCB-1 material strongly indicated that the intralayer structure was well-preserved under the mild condition. However, a nonaqueous fluoride/chloride solution was needed in the delamination of the PREFER lamellar zeolite, and dimethylformamide was proved to be a suitable solvent in producing the delaminated UCB-2 material [66].

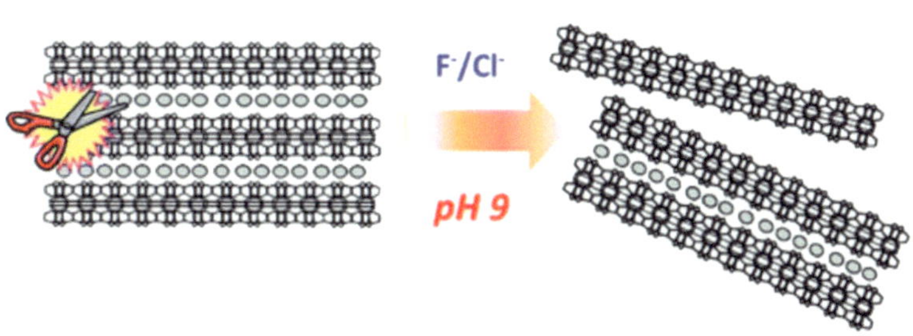

Figure 6.
Mild delamination of MWW-layered zeolite with the assistance of F^- and Cl^- ions.

A more simple and mild delamination strategy without surfactant molecules and sonication was also proposed by Zones and Katz via the isomorphous substitution of Al for B in the B-containing MWW zeolite using aqueous $Al(NO_3)_3$ solution (**Figure** 7) [67]. The presence of neutral amine SDA and the careful control of temperature were proved to be critical for the success of this one-step delamination process. The obtained delaminated Al-containing ERB-1-del showed a similar morphology of single thin layers as the conventional delaminated ITQ-2 material and showed higher activity in Friedel-Crafts acylation reactions involving bulky substrates compared to 3D ERB-1C zeolite, due to the enhanced external surface area (133 *vs* 53 m^2/g). An all-silica analog of ERB-1-del can be obtained by treating ERB-1 lamellar precursor with warm $Zn(NO_3)_2$ solution with a pH value of ~1 [68]. However, the one-step mild delamination strategy is now restricted to ERB-1 lamellar precursor and has its limitations in the general application.

3.2 Interlayer expansion with large-sized silane agents assisted by deconstruction-reconstruction

Although the simple high-temperature acid treatment is able to give interlayer-expanded structures, the assistant of silane agents helps the formation of more ordered and stable structures [69, 70]. The interlayer space of most lamellar precursors is large enough for monomolecular silane agent to insert two additional Si atoms in the pore window. To introduce more Si atoms and construct larger interlayer pore structures, the interlayer silylation achieved by large-sized silane molecules should be performed over the swollen intermediate [71, 72]. However, the conventional swelling process under the basic condition induces severe amorphization of the layer structures. To tailor the interlayer space, an interlayer deconstruction-reconstruction strategy was proposed by Wu et al. [73]. The PLS-3 lamellar precursor was firstly deconstructed by a mild acid treatment to induce the partial removal of SDA molecules and then reconstructed in the aqueous solution

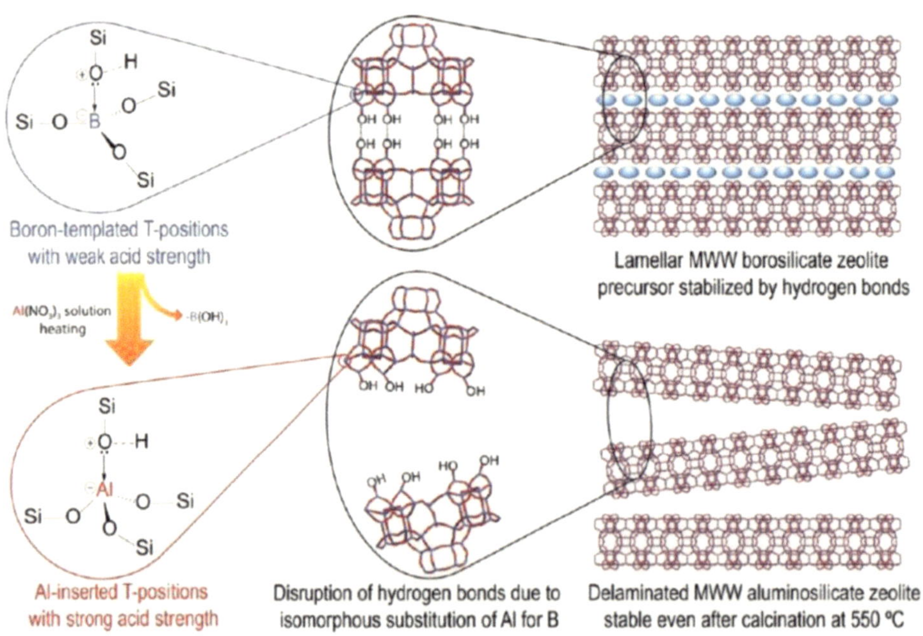

Figure 7.
Single-step delamination of MWW-layered zeolite via the isomorphous substitution of Al for B.

containing bulky ammonium molecules, resulting in a lamellar precursor ECNU-9(P) with an enlarged interlayer space (**Figure 8**). The bulky ammonium molecules are the classical SDA for the PREFER-layered zeolite having the same layer structure as PLS-3. Thus, the layer structure of PLS-3 was well-protected in the reconstruction process. The interlayer-expanded ECNU-9 zeolite with large/extra-large pore system of 14 × 12-R was synthesized by the intercalation of a bulky single four ring-shaped silane agent into ECNU-9(P). Ti-ECNU-9, obtained by incorporating Ti atoms into ECNU-9 via H_2TiF_6 treatment, exhibited significantly high activity in the epoxidation reaction of cyclohexene, compared with 3D Ti-FER and the one expanded by monomolecular silane agent.

A similar strategy was also reported in the silylation of HUS-2-layered zeolite [74]. A distance of a half unit cell along [001] the direction for the up-and-down silanols on the layer surface and the presence of silicon vacancy induced the structural collapse of HUS-2 upon calcination [28]. Thus, a dimeric silane agent with two active groups attached to each silicon atom was applied to connect the relative long-distance up-and-down silanols and simultaneously fill the vacancy. The original HUS-2 lamellar zeolite could not offer enough interlayer space to accommodate the bulky dimeric silane agent. Thus, a Sub-HUS-2 material with interlayer deconstructed structure was firstly prepared and then reconstructed with the assistance of bulky ammonium molecules, resulting in an enlarged interlayer space for the following silylation. The obtained interlayer-expanded ECNU-19 zeolite with intersecting 12 × 8-R pore system was confirmed by the structure refinement.

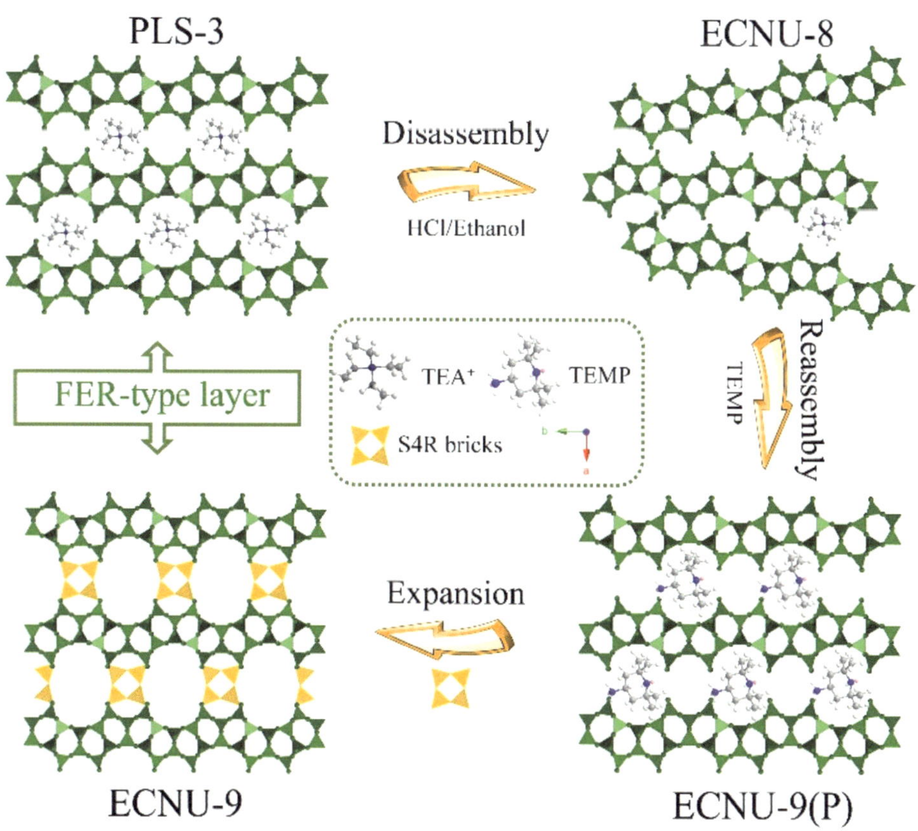

Figure 8.
Scheme description of interlayer-expanded ECNU-9 zeolite via deconstruction-reconstruction strategy.

3.3 Reorganization of layer stacking by rapid dissolution-recrystallization method

The reorganization of layer-stacking style is a useful strategy to produce partial delaminated materials, which preserved more micropores compared to full delaminated ones [34, 75]. Although the delaminated analogs can be obtained by direct hydrothermal synthesis [10, 76, 77], the posttreatment modification is easier to control. Partial detemplating by a mild acid treatment over MWW-layered zeolite produced MCM-56 with higher external surface area than conventional 3D MWW zeolite [34, 75]. In addition to this, a novel strategy of rapid dissolution-recrystallization was reported to give an intergrowth ECNU-5 zeolite with two different polymorphs of ECNU-5A and ECNU-5B (**Figure 9**) [12]. In the SDA-containing alkaline solution, the 3D all-silica MWW precursor ITQ-1 was fast dissolved to silica species in 1 h, which were then recrystallized to a lamellar material ECNU-5P with the assistant of the unique SDA of 1,3-bis(cyclohexyl) imidazolium hydroxide. The asymmetrical cyclohexyl groups in the SDA molecule were responsible for the disordered stacking style and the resultant intergrowth structure by geometry mismatching. A new pore window with the size close to 14R was found in the ECNU-5 zeolite, different from the 12R cage in the traditional 3D MWW because of the misconnection of the up-and-down pockets from neighboring layers. ECNU-5 zeolite was later proved to be analogous to SSZ-70 zeolite [11, 77].

A direct synthesized swollen MWW material ECNU-7P can be obtained by introducing the surfactant of cetyltrimethylammonium bromide (CTAB) in the above recrystallization process (**Figure 10**) [78]. The strong inorganic-organic interaction between the surfactant and structure layer, proved by 2D ^{1}H—^{29}Si solid-state NMR, induced the formation of the resultant alternating stacking of inorganic MWW layers and the organic CTAB layers. The calcined ECNU-7 zeolite exhibited a hierarchical pore system and larger external surface area than that of 3D MWW zeolite. Heteroatoms, like Al and Ti, can be isomorphously incorporated into the ECNU-7 zeolite, and they showed a superior catalytic activity in the cracking of 1,3,5-triisopropylbenzene and the epoxidation of propylene with bulky cumene hydroperoxide as oxidant, respectively, compared to the traditional 3D MWW zeolite [79].

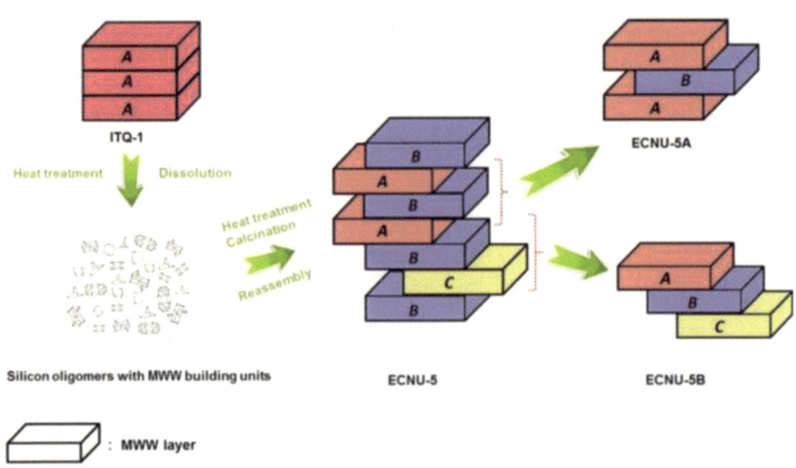

Figure 9.
Scheme description of intergrowth ECNU-5 zeolite with novel MWW layer arrangement by rapid dissolution-recrystallization method.

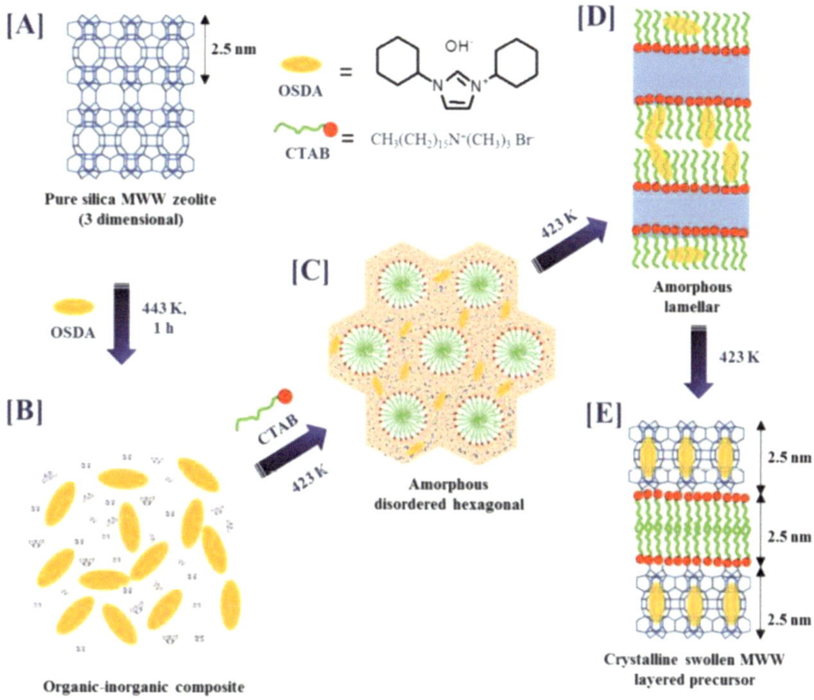

Figure 10.
Direct synthesis of swollen ECNU-7P zeolite.

4. Conclusions

Layered zeolites, composed of zeolitic nanosheets linked by hydrogen bonds, provide modifiable precursors for the synthesis of novel zeolite derivatives with open pore system or large external surface area. The newly developed synthesis strategies, including the usage of delicately designed bifunctional amphiphilic SDAs and the top-down selective hydrolysis of 3D germanosilicates, have unexpectedly broken the limitation of conventional hydrothermal synthesis and largely expanded the family of layered zeolites. The successful example of layered MFI zeolite synthesized by bifunctional SDA indicated the potential of design synthesis and encouraged the researchers to explore more novel-layered zeolites. The selective degradation of germanosilicates, transforming the 3D zeolite to 2D lamellar zeolite, is highly related to the orientation distribution of instable bonds in the framework. To explore other zeolite structures besides germanosilicates with such regularity is expected to create new layered precursors.

On the other hand, several improved modification procedures have been proposed to simplify the multistep conventional processes and avoid the harsh treatment conditions. The variation of post-synthesis modifications is potential in providing more derivatives for the layered precursors, showing the infinite possibilities of the modifiable 2D layered zeolites.

Acknowledgements

The authors gratefully acknowledge the financial supports from the NSFC of China (21872025, 21533002, and 21603075) and China Ministry of Science and Technology (2016YFA0202804).

Conflict of interest

The authors declare no conflict of interest.

Author details

Hao Xu and Peng Wu*
Shanghai Key Laboratory of Green Chemistry and Chemical Processes, School of Chemistry and Molecular Engineering, East China Normal University, Shanghai, China

*Address all correspondence to: pwu@chem.ecnu.edu.cn

References

[1] Baerlocher CH, McCusker LB. Database of Zeolite Structures. 2019. Available from: http://www.iza-structure.org/databases/

[2] Díaz U, Corma A. Layered zeolitic materials: An approach to designing versatile functional solids. Dalton Transactions. 2014;**43**:10292-10316. DOI: 10.1039/c3dt53181c

[3] Choi M, Na K, Kim J, Sakamoto Y, Terasaki O, Ryoo R. Stable single-unit-cell nanosheets of zeolite MFI as active and long-lived catalysts. Nature. 2009;**461**:246-250. DOI: 10.1038/nature08288

[4] Roth WJ, Shvets OV, Shamzhy M, Chlubná P, Kubů M, Nachtiga P, et al. Postsynthesis transformation of three-dimensional framework into a lamellar zeolite with modifiable architecture. Journal of the American Chemical Society. 2011;**133**:6130-6133. DOI: 10.1021/ja200741r

[5] Leonowicz ME, Lawton JA, Lawton SL, Rubin MK. MCM-22: A molecular sieve with two independent multidimensional channel systems. Science. 1994;**264**:1910-1913. DOI: 10.1126/science.264.5167.1910

[6] Puppe L, Weisser J. Crystalline aluminosilicate PSH-3 and its process of preparation. U.S. Patent 4439409; 1984

[7] Zones SI, Hwang S, Davis ME. Studies of the synthesis of SSZ-25 zeolite in a "mixed-template" system. Chemistry-A European Journal. 2001;**7**:1990-2001. DOI: 10.1002/1521-3765(20010504)7:9<1990::AID-CHEM1990>3.0.CO;2-G

[8] Camblor MA, Corma A, Díaz-Cabañas M. Synthesis and structural characterization of MWW type zeolite ITQ-1, the pure silica analog of MCM-22 and SSZ-25. The Journal of Physical Chemistry B. 1998;**102**:44-51. DOI: 10.1021/jp972319k

[9] Millini R, Perego G, Parker WO Jr, Bellussi G, Carluccio L. Layered structure of ERB-1 microporous borosilicate precursor and its intercalation properties towards polar molecules. Microporous Materials. 1995;**4**:221-230. DOI: 10.1016/0927-6513(95)00013-Y

[10] Roth WJ, Dorset DL, Kennedy GJ. Discovery of new MWW family zeolite EMM-10: Identification of EMM-10P as the missing MWW precursor with disordered layers. Microporous and Mesoporous Materials. 2011;**142**:168-177. DOI: 10.1016/j.micromeso.2010.10.052

[11] Smeets S, Berkson ZJ, Xie D, Zones SI, Wan W, Zou X, et al. Well-defined silanols in the structure of the calcined high-silica zeolite SSZ-70: New understanding of a successful catalytic material. Journal of the American Chemical Society. 2017;**139**:16803-16812. DOI: 10.1021/jacs.7b08810

[12] Xu L, Ji X, Jiang J, Han L, Che S, Wu P. Intergrown zeolite MWW polymorphs prepared by the rapid dissolution-recrystallization route. Chemistry of Materials. 2015;**27**:7852-7860. DOI: 10.1021/acs.chemmater.5b03658

[13] Schreyeck L, Caullet P, Mougenel JC, Guth JL, Marler B. PREFER: A new layered (alumino) silicate precursor of FER-type zeolite. Microporous Materials. 1996;**6**:259-271. DOI: 10.1016/0927-6513(96)00032-6

[14] Ikeda T, Kayamori S, Mizukami F. Synthesis and crystal structure of layered silicate PLS-3 and PLS-4 as a topotactic zeolite precursor. Journal of Materials Chemistry. 2009;**19**:5518-5525. DOI: 10.1039/B905415D

[15] Burton A, Accardi RJ, Lobo RF, Falcioni M, Deem MW. MCM-47: A highly crystalline silicate composed of hydrogen-bonded ferrierite layers. Chemistry of Materials. 2000;**12**: 2936-2942. DOI: 10.1021/cm000243q

[16] Dorset DL, Kennedy GJ. Crystal structure of MCM-65: An alternative linkage of ferrierite layers. The Journal of Physical Chemistry B. 2004;**108**:15216-15222. DOI: 10.1021/jp040305q

[17] Komura K, Murase T, Sugi Y, Koketsu M. Synthesis of boron-containing CDS-1 zeolite by topotactic dehydration-condensation of [B]-PLS-1 prepared from layered silicate H-LDS. Chemistry Letters. 2010;**39**: 948-949. DOI: 10.1246/cl.2010.948

[18] Gies H, Müller U, Yilmaz B, Feyen M, Tatsumi T, Imai H, et al. Interlayer expansion of the hydrous layer silicate RUB-36 to a functionalized, microporous framework silicate: Crystal structure analysis and physical and chemical characterization. Chemistry of Materials. 2012;**24**:1536-1545. DOI: 10.1021/cm300525u

[19] Blake AJ, Franklin KR, Lowe BM. Preparation and properties of piperazine silicate (EU-19) and a silica polymorph (EU-20). Journal of the Chemical Society Dalton Transactions. 1988;**10**:2513-2517. DOI: 10.1039/DT9880002513

[20] Zanardi S, Alberti A, Cruciani G, Corma A, Fornes V, Brunelli M. Crystal structure determination of zeolite nu-6(2) and its layered precursor nu-6(1). Angewandte Chemie, International Edition. 2004;**43**:4933-4937. DOI: 10.1002/anie.200460085

[21] Oberhagemann U, Bayat P, Marler B, Gies H, Rius J. A layer silicate: Synthesis and structure of the zeolite precursor RUB-15—$[N(CH_3)_4]_8[Si_{24}O_{52}(OH)_4]\cdot 20H_2O$. Angewandte Chemie International Edition in English. 1996;**35**:2869-2872. DOI: 10.1002/anie.199628691

[22] Marler B, Ströter N, Gies H. The structure of the new pure silica zeolite RUB-24, $Si_{32}O_{64}$, obtained by topotactic condensation of the intercalated layer silicate RUB-18. Microporous and Mesoporous Materials. 2005;**83**:201-211. DOI: 10.1016/j.micromeso.2005.04.007

[23] Wang YX, Gies H, Lin JH. Crystal structure of the new layer silicate RUB-39 and its topotactic condensation to a microporous zeolite with framework type RRO. Chemistry of Materials. 2007;**19**:4181-4188. DOI: 10.1021/cm0706907

[24] Wheatley PS, Morris RE. Calcination of a layered aluminofluorophosphate precursor to form the zeolitic AFO framework. Journal of Materials Chemistry. 2006;**16**:1035-1037. DOI: 10.1039/B518265D

[25] Morris AA, Bignami GPM, Tian Y, Navarro M, Firth DS, Cejka J, et al. In situ solid-state NMR and XRD studies of the ADOR process and the unusual structure of zeolite IPC-6. Nature Chemistry. 2017;**9**:1012-1018. DOI: 10.17630/47861a6e-f7c4-466e-9108-d4adb104f74d

[26] Liu X, Mao W, Jiang J, Lu X, Peng M, Xu H, et al. Topotactic conversion of alkali-treated intergrown germanosilicate CIT-13 into single-crystalline ECNU-21 zeolite as shape-selective catalyst for ethylene oxide hydration. Chemistry-A European Journal. 2019;**25**:4520-4529. DOI: 10.1002/chem.201900173

[27] Rollmann LD, Schlenker JL, Lawton SL, Kennedy CL, Kennedy GJ. MCM-69, a novel layered analogue of EU-19. Microporous and Mesoporous Materials. 2002;**53**:179-193. DOI: 10.1016/S1387-1811(02)00338-4

[28] Tsunoji N, Ikeda T, Ide Y, Sadakane M, Sano T. Synthesis and characteristics of novel layered silicates HUS-2 and HUS-3 derived from a SiO_2-choline hydroxide-NaOH-H_2O system. Journal of Materials Chemistry. 2012;**22**: 13682-13690. DOI: 10.1039/C2JM31872E

[29] Maheshwari S, Jordan E, Kumar S, Bates FS, Penn RL, Shantz DF, et al. Layer structure preservation during swelling, pillaring, and exfoliation of a zeolite precursor. Journal of the American Chemical Society. 2008;**130**:1507-1516. DOI: 10.1021/ja077711i

[30] Corma A, Fornes V, Pergher SB, Maesen TLM, Buglass JG. Delaminated zeolite precursors as selective acidic catalysts. Nature. 1998;**396**:353-356. DOI: 10.1038/24592

[31] Zukal A, Dominguez I, Mayerova J, Cejka J. Functionalization of delaminated zeolite ITQ-6 for the adsorption of carbon dioxide. Langmuir. 2009;**25**:10314-10321. DOI: 10.1021/la901156z

[32] Roth WJ, Kresgea CT, Vartuli JC, Leonowicz ME, Fung AS, McCullen SB. MCM-36: The first pillared molecular sieve with zeolite properties. Studies in Surface Science and Catalysis. 1995;**94**:301-308. DOI: 10.1016/S0167-2991(06)81236-X

[33] Wu P, Ruan J, Wang L, Wu L, Wang Y, Liu Y, et al. Methodology for synthesizing crystalline metallosilicates with expanded pore windows through molecular alkoxysilylation of zeolitic lamellar precursors. Journal of the American Chemical Society. 2008;**130**:8178-8187. DOI: 10.1021/ja0758739

[34] Wang Y, Liu Y, Wang L, Wu H, Li X, He M, et al. Postsynthesis, characterization, and catalytic properties of aluminosilicates analogous to MCM-56. The Journal of Physical Chemistry C. 2009;**113**:18753-18760. DOI: 10.1021/jp904436c

[35] Roth WJ, Čejka J. Two-dimensional zeolites: Dream or reality? Catalysis Science & Technology. 2011;**1**:43-53. DOI: 10.1039/c0cy00027b

[36] Roth WJ, Gil B, Makowski W, Marszalek B, Eliášová P. Layer like porous materials with hierarchical structure. Chemical Society Reviews. 2016;**45**:3400-3438. DOI: 10.1039/c5cs00508f

[37] Schwanke A, Pergher S. Lamellar MWW-type zeolites: Toward elegant nanoporous materials. Applied Sciences. 2018;**8**:1636-1650. DOI: 10.3390/app8091636

[38] Xu L, Wu P. Diversity of layered zeolites: From synthesis to structural modifications. New Journal of Chemistry. 2016;**40**:3968-3981. DOI: 10.1039/c5nj02829a

[39] Opanasenko MV, Roth WJ, Čejka J. Two-dimensional zeolites in catalysis: Current status and perspectives. Catalysis Science & Technology. 2016;**6**:2467-2484. DOI: 10.1039/c5cy02079d

[40] Wu P, Xu H, Xu L, Liu Y, He M. MWW-type Titanosilicate: Synthesis, Structural Modification and Catalytic Applications to Green Oxidation. United States: Springer; 2013. DOI: 10.1007/978-3-642-39115-6

[41] Na K, Choi M, Park W, Sakamoto Y, Terasaki O, Ryoo R. Pillared MFI zeolite nanosheets of a single-unit-cell thickness. Journal of the American Chemical Society. 2010;**132**:4169-4177. DOI: 10.1021/ja908382n

[42] Wang J, Xu L, Zhang K, Peng H, Wu H, Jiang JG, et al. Multilayer structured MFI-type titanosilicate: Synthesis and catalytic properties in selective epoxidation of bulky molecules. Journal

of Catalysis. 2012;**288**:16-23. DOI: 10.1016/j.jcat.2011.12.023

[43] Xu L, Li CG, Zhang K, Wu P. Bifunctional tandem catalysis on multilamellar organic-inorganic hybrid zeolites. ACS Catalysis. 2014;**4**: 2959-2968. DOI: 10.1021/cs500653p

[44] Park W, Yu D, Na K, Jelfs KE, Slater B, Sakamoto Y, et al. Hierarchically structure-directing effect of multi-ammonium surfactants for the generation of MFI zeolite nanosheets. Chemistry of Materials. 2011;**23**: 5131-5137. DOI: 10.1021/cm201709q

[45] Na K, Jo C, Kim J, Cho K, Jung J, Seo Y, et al. Directing zeolite structures into hierarchically nanoporous architectures. Science. 2011;**333**:328-332. DOI: 10.1126/science.1204452

[46] Xu D, Ma Y, Jing Z, Han L, Singh B, Feng J, et al. π-π interaction of aromatic groups in amphiphilic molecules directing for single-crystalline mesostructured zeolite nanosheets. Nature Communications. 2014;**5**: 4262-4270. DOI: 10.1038/ncomms5262

[47] Singh BK, Xu D, Han L, Ding J, Wang Y, Che S. Synthesis of single-crystalline mesoporous ZSM-5 with three-dimensional pores via the self-assembly of a designed triply branched cationic surfactant. Chemistry of Materials. 2014;**26**:7183-7188. DOI: 10.1021/cm503919h

[48] Shen X, Mao W, Ma Y, Xu D, Wu P, Terasaki O, et al. A hierarchical MFI zeolite with a two-dimensional square mesostructure. Angewandte Chemie. 2018;**130**:732-736. DOI: 10.1002/ange.201710748

[49] Zhang Y, Ma Y, Che S. Synthesis of lamellar mesostructured ZSM-48 nanosheets. Chemistry of Materials. 2018;**30**:1839-1843. DOI: 10.1021/acs.chemmater.8b00146

[50] Seo Y, Lee S, Jo C, Ryoo R. Microporous aluminophosphate nanosheets and their nanomorphic zeolite analogues tailored by hierarchical structure-directing amines. Journal of the American Chemical Society. 2013;**135**:8806-8809. DOI: 10.1021/ja403580j

[51] Luo HY, Michaelis VK, Hodges S, Griffin RG, Román-Leshkov Y. One-pot synthesis of MWW zeolite nanosheets using a rationally designed organic structure-directing agent. Chemical Science. 2015;**6**:6320-6324. DOI: 10.1039/C5SC01912E

[52] Margarit VJ, Martínez-Armero ME, Navarro MT, Martínez C, Corma A. Direct dual-template synthesis of MWW zeolite monolayers. Angewandte Chemie, International Edition. 2015;**54**:13724-13728. DOI: 10.1002/anie.201506822

[53] Sun J, Bonneau C, Cantín Á, Corma A, Díaz-Cabañas MJ, Moliner M, et al. The ITQ-37 mesoporous chiral zeolite. Nature. 2009;**458**:1154-1158. DOI: 10.1038/nature07957

[54] Jiang J, Jorda JL, Yu J, Baumes LA, Mugnaioli E, Diaz-Cabanas MJ, et al. Synthesis and structure determination of the hierarchical meso-microporous zeolite ITQ-43. Science. 2011;**333**: 1131-1134. DOI: 10.1126/science.1208652

[55] Petkov PS, Aleksandrov HA, Valtchev V, Vayssilov GN. Framework stability of heteroatom-substituted forms of extra-large-pore Ge-silicate molecular sieves: The case of ITQ-44. Chemistry of Materials. 2012;**24**: 2509-2518. DOI: 10.1021/cm300861e

[56] Mazur M, Wheatley PS, Navarro M, Roth WJ, Položij M, Mayoral A, et al. Synthesis of 'unfeasible' zeolites. Nature Chemistry. 2015;**8**:58-62. DOI: 10.1038/nchem.2374

[57] Wheatley PS, Chlubná-Elišová P, Greer H, Zhou W, Seymour VR,

Dawson DM, et al. Zeolites with continuously tuneable porosity. Angewandte Chemie, International Edition. 2014;**53**:13210-13214. DOI: 10.1002/anie.201407676

[58] Roth WJ, Nachtigall P, Morris RE, Wheatley PS, Seymour VR, Ashbrook SE, et al. A family of zeolites with controlled pore size prepared using a top-down method. Nature Chemistry. 2013;**5**:628-633. DOI: 10.1038/nchem.1662

[59] Verheyen E, Joos L, Van Havenbergh K, Breynaert E, Kasian N, Gobechiya E, et al. Design of zeolite by inverse sigma transformation. Nature Materials. 2012;**11**:1059. DOI: 10.1038/nmat3455

[60] Shamzhy M, Opanasenko M, Tian Y, Konysheva K, Shvets O, Morris RE, et al. Germanosilicate precursors of ADORable zeolites obtained by disassembly of ITH, ITR, and IWR zeolites. Chemistry of Materials. 2014;**26**:5789-5798. DOI: 10.1021/cm502953s

[61] Chlubná-Eliášová P, Tian Y, Pinar AB, Kubů M, Čejka J, Morris RE. The assembly-disassembly-organization-reassembly mechanism for 3D-2D-3D transformation of germanosilicate IWW zeolite. Angewandte Chemie, International Edition. 2014;**53**:7048-7052. DOI: 10.1002/anie.201400600

[62] Kasneryk V, Shamzhy M, Opanasenko M, Wheatley PS, Morris SA, Russell SE, et al. Expansion of the ADOR strategy for the synthesis of zeolites: The synthesis of IPC-12 from zeolite UOV. Angewandte Chemie, International Edition. 2017;**56**:4324-4327. DOI: 10.1002/anie.201700590

[63] Firth DS, Morris SA, Wheatley PS, Russell SE, Slawin AMZ, Dawson DM, et al. Assembly-disassembly-organization-reassembly synthesis of zeolites based on cfi-type layers. Chemistry of Materials. 2017;**29**:5605-5611. DOI: 10.1021/acs.chemmater.7b01181

[64] Kasian N, Tuel A, Verheyen E, Kirschhock CEA, Taulelle F, Martens JA. NMR evidence for specific germanium siting in IM-12 zeolite. Chemistry of Materials. 2014;**26**:5556-5565. DOI: 10.1021/cm502525w

[65] Ogino I, Nigra MM, Hwang S, Ha J, Rea T, Zones SI, et al. Delamination of layered zeolite precursors under mild conditions: Synthesis of UCB-1 via fluoride/chloride anion-promoted exfoliation. Journal of the American Chemical Society. 2011;**133**:3288-3291. DOI: 10.1021/ja111147z

[66] Eilertsen EA, Ogino I, Hwang S, Rea T, Yeh S, Zones SI, et al. Nonaqueous fluoride/chloride anion-promoted delamination of layered zeolite precursors: Synthesis and characterization of UCB-2. Chemistry of Materials. 2011;**23**:5404-5408. DOI: 10.1021/cm202364q

[67] Ouyang X, Hwang S, Runnebaum RC, Dan X, Wanglee Y, Rea T, et al. Single-step delamination of a MWW borosilicate layered zeolite precursor under mild conditions without surfactant and sonication. Journal of the American Chemical Society. 2014;**136**:1449-1461. DOI: 10.1021/ja410141u

[68] Ouyang X, Wanglee Y, Hwang S, Xie D, Rea T, Zones SI, et al. Novel surfactant-free route to delaminated all-silica and titanosilicate zeolites derived from a layered borosilicate MWW precursor. Dalton Transactions. 2014;**43**:10417-10429. DOI: 10.1039/c4dt00383g

[69] Xu H, Yang B, Jiang J, Jia L, He M, Wu P. Post-synthesis and adsorption properties of interlayer-expanded PLS-4 zeolite. Microporous and Mesoporous Materials. 2013;**169**:88-96. DOI: 10.1016/j.micromeso.2012.10.005

[70] Ikeda T, Kayamori S, Oumi Y, Mizukami F. Structure analysis of Si-atom pillared lamellar silicates having micropore structure by powder X-ray diffraction. The Journal of Physical Chemistry C. 2010;**114**:3466-3476. DOI: 10.1021/jp912026n

[71] Corma A, Díaz U, García T, Sastre G, Velty A. Multifunctional hybrid organic-inorganic catalytic materials with a hierarchical system of well-defined micro- and mesopores. Journal of the American Chemical Society. 2010;**132**:15011-15021. DOI: 10.1021/ja106272z

[72] Xu H, Fu L, Jiang J, He M, Wu P. Preparation of hierarchical MWW-type titanosilicate by interlayer silylation with dimeric silane. Microporous and Mesoporous Materials. 2014;**189**:41-48. DOI: 10.1016/j.micromeso.2013.09.041

[73] Yang B, Jiang J, Xu H, Wu H, He M, Wu P. Synthesis of extra-large-pore zeolite ECNU-9 with intersecting 14*12-ring channels. Angewandte Chemie, International Edition. 2018;**57**:9515-9519. DOI: 10.1002/anie.201805535

[74] Yang B, Jiang J, Xu H, Wu H, Wu P. Synthesis of large-pore ECNU-19 material (12×8-R) via interlayer-expansion of HUS-2 lamellar silicate. Chinese Journal of Chemistry. 2018;**36**:227-232. DOI: 10.1002/cjoc.201700607

[75] Liu G, Jiang J, Yang B, Fang X, Xu H, Peng H, et al. Hydrothermal synthesis of MWW-type stannosilicate and its post-structural transformation to MCM-56 analogue. Microporous and Mesoporous Materials. 2013;**165**:210-218. DOI: 10.1016/j.micromeso.2012.08.025

[76] Corma A, Diaz U, Fornés V, Guil JM, Martínez-Triguero J, Creyghton EJ. Characterization and catalytic activity of MCM-22 and MCM-56 compared with ITQ-2. Journal of Catalysis. 2000;**191**:218-224. DOI: 10.1006/jcat.1999.2774

[77] Archer RH, Carpenter JR, Hwang S, Burton AW, Chen C, Zones SI, et al. Physicochemical properties and catalytic behavior of the molecular sieve SSZ-70. Journal of the American Chemical Society. 2010;**22**:2563-2572. DOI: 10.1021/cm9035677

[78] Xu L, Ji X, Li S, Zhou Z, Du X, Sun J, et al. Self-assembly of cetyltrimethylammonium bromide and lamellar zeolite precursor for the preparation of hierarchical MWW zeolite. Chemistry of Materials. 2016;**28**:4512-4521. DOI: 10.1021/acs.chemmater.6b02155

[79] Ji X, Xu L, Du X, Lu X, Lu W, Sun J, et al. Simple CTAB surfactant-assisted hierarchical lamellar MWW titanosilicate: A high-performance catalyst for selective oxidations involving bulky substrates. Catalysis Science & Technology. 2017;7: 2874-2885. DOI: 10.1039/c7cy00756f

Section 3

Application of Zeolite Materials

Chapter 6

Zeolites: An Emerging Material for Gas Storage and Separation Applications

Nandini Das and Jugal Kishore Das

Abstract

Zeolites are one of the amazing materials available in nature because of their structural pores. Interestingly, these god-gifted properties of zeolite can be used in gas separation and storage application. Actually, hydrogen separation and its storage are now a thrust research area. Hydrogen is considered as a 'clean energy,' which is indispensable for global affluence and alternative energy for future. But hydrogen is not accessible in its pure form during the industrial synthesis process and comes out with some other impurities like CO_2 (GHG) and other gases. So, the production of carbon-free hydrogen and its storage is so much vital. In conventional technologies, few concerns are always existed during gas separation and also in storage process. Recently, membrane-based separation process is a highly demanding technology in the industry and shows some advantages as compared to conventional process. Based on this concept, in this chapter, three different types of zeolites, that is, DDR, SAPO 34, and Bikitaite are highlighted. Here, we described the advanced synthesis process and the mechanism towards the development of high-quality nearly defect-free membranes on cheaper support. Finally, the evaluation of membranes is described through gas permeation and selectivity results of different single gas and mixture gas composition. In addition, storage capacity of H_2 by zeolite/surface-modified zeolites is included in this chapter.

Keywords: zeolite membrane, gas separation, gas storage, clean energy, pollution control

1. Introduction

Gas separation and storage processes are essentially important to various aspects in human society, such as energy consumption, environmental security, and industrial production. Energy and environmental concerns are currently at the forefront of global attention. So, carbon dioxide separation is crucial to the mitigation of greenhouse effect [1–3]. Besides, separation of hydrogen and methane together with storage is indispensable for the prevalent use of clean energy. In the case of toxic gases, the separation and storage of ammonia and carbon monoxide are important for pollution control and the synthesis of industrial chemicals. The conventional gas separation technologies such as pressure swing adsorption (PSA), cryogenic distillation, etc. are very energy intensive as well as capital intensive. Also separation methods like liquid adsorbent are cost-effective. In the distillation process, the repeated evaporating-condensing cycle of the mixture under harsh conditions is a problematical job. Also generation of liquid adsorbent is a main concern which required the heating and cooling of massive solvent medium to release adsorbed gas. Due to these negative aspects, the potential of emerging technologies based on adsorption or membrane separations is highly amiable alternative process and has been proposed as more energy-efficient technologies [4–8]. According to existing literatures, membrane-based separation technology only consumes 10% energy of that for distillation [9]. From the industrial perspective of storage and separation of different gases, adsorption-based technique is more amicable and commendable due to its superiority to other techniques like simplicity of design, easy operation, and low cost. The separation efficiency relies on internal porosity and surface properties of solid adsorbent due to their key role in gas sorption. Alternatively, molecular properties of the adsorbent such as chemical affinity or molecular size of the separated components play a vital role in the separation process. Separation and purification, meanwhile, involve the selective adsorption of particular species from gas mixtures. Also, gas storage requires elevated pressures due to volumetric capacity considerations of porous materials and the need to deliver gas at ambient pressure or above.

Nanoporous materials have attracted huge interest among the communities of materials science, chemical engineering, and chemistry due to their excellent properties such as high surface area, large pore volume, and specific surface chemistry. The term *nanoporous* refers to any material with a pore size below ~100 nm. According to the International Union of Pure and Applied Chemistry (IUPAC) guidelines, nanoporous materials encompass both the microporous (<2 nm) and mesoporous (2–50 nm) regimes. As per literature data, porous materials like zeolites, carbon, aluminophosphates, carbon nanotubes, silica gel, pillared clays, inorganic and polymeric resins, MOFs, and MOFs composites have been investigated as adsorbents. In industry some of the adsorbents are now used for different applications. In the literature, relevant reviews and monographs have discussed the syntheses process, characterizations, and the adsorption properties of these porous materials [9–14]. The importance of porous materials for different application is summarized in the literature, which can be helpful for the next-generation researcher. Among all porous material, especially in the microporous family, zeolites are the first and foremost emerging materials and attracted increasing interest because of their unique physical and chemical properties, such as high surface area, high chemical resistance, extraordinary mechanical properties, good adsorption, and catalytic properties due to specific surface chemistry [15–22]. These peculiar and amazing properties have highlighted the potential of this material in a variety of applications and particularly in the area of gas separation and storage application [23–34]. Zeolites are traditionally referred to as a family of open-framework aluminosilicate materials consisting of orderly distributed micropores in molecular dimensions. Topologically, zeolites are three-dimensional networks of corner-sharing tetrahedral TO_4 ("T" denotes tetrahedrally coordinated Si, Al, or P), and different ways of tetrahedra connection lead to a diversity of zeolite framework types based on various compositions [35]. Silica zeolites consist of four-coordinated Si bridged by oxygen atoms [36]. To date, 235 distinct zeolite framework types have been identified in natural or synthetic zeolites, each of which has been assigned a three-letter code by the International Zeolite Association (**Figure 1**) [37].

For zeolite synthesis, the well-known conventional hydrothermal method is a widely used technique. Besides the other synthesis method like sonochemical and sonochemical-assisted hydrothermal method, microwave-assisted methods are more popular and advanced synthesis process to achieve phase pure high-quality zeolites in terms of their shape, size, porosity, uniform structure, and better crystallinity [38]. Furthermore, for the synthesis of zeolite membrane on the porous support, the in-situ and ex-situ (secondary growth) hydrothermal techniques are

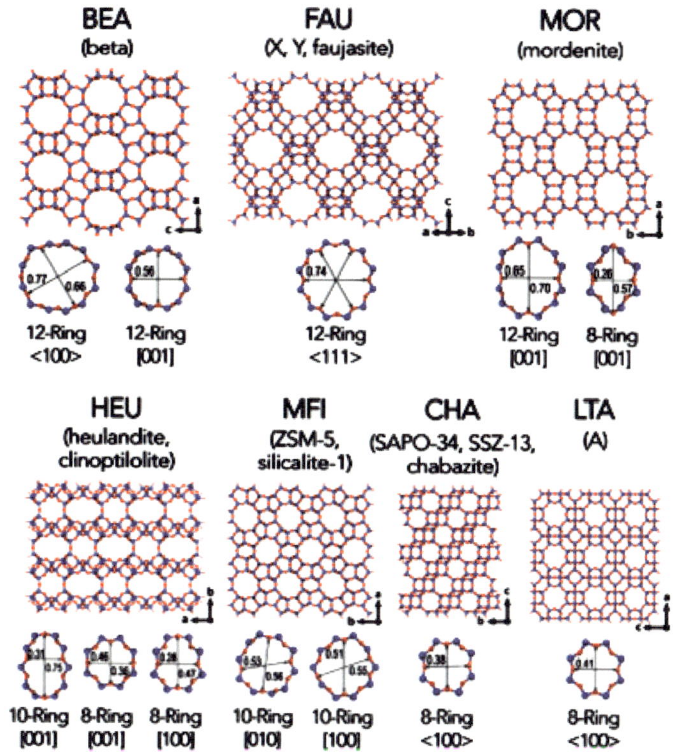

Figure 1.
Framework types of different zeolites [36, 37].

the well known process and more popular among other synthesis routes. In the case of in situ hydrothermal process, the porous support is immersed into the synthesis solution, and the membrane layer is formed directly through direct crystallization in a suitable time period. But in this process, the probability of attaining the high-quality membrane on the support is less. So ex situ hydrothermal method which is also known as seeded growth technique is an effective and accepted approach towards the development of better membrane on the support surface. This method has potential advantages in terms of achieved preferential orientation control of membrane microstructure and higher reproducibility if compared with the in situ synthesis method [39]. Extensive studies have been performed aiming at investigating potential of zeolites and derived membranes for gas separation [40–56]. Specific zeolites have a high capacity and selectivity for the gases of interest, leading to compact and efficient separation/storage systems.

To make this book chapter comprehensive, here three different types of zeolitic material were focused. The first one is siliceous deca-dodecasil 3R (DDR) zeolite which has elliptical pore openings defined by 8-member ring windows with an effective size of 0.36 × 0.44 nm^2, and it is useful for separation of small-sized gas [27, 43, 50]. Another important zeolite is silicoaluminophosphate (SAPO 34). SAPO 34, a chabazite zeolite with a composition of $Si_xAl_yP_zO_2$, where x = 0.01–0.98, y = 0.01–0.60, and z = 0.01–0.52, has an average pore size of 0.38 nm and plays an important role for gas separation application [32]. The last one is Bikitaite (BIK) zeolite having a unit cell chemical composition $Li_2(Al_2Si_4O_{12})$-$2H_2O$ [11, 36]. It is a small pore (diameter 0.28–0.37 nm) sized zeolite and has been studied for various applications and most notably has shown better performance in gas separation and storage application. The detail synthesis protocol and techniques used to synthesize zeolites and high-quality membrane have been discussed here.

After all, the performance of the developed materials was discussed elaborately for better understanding and presents the future aspect of these materials.

2. Syntheses of three different types of zeolitic material

2.1 Materials

The chemical reagents used are boehmite powder (SASOL, Germany), colloidal silica (Ludox HS- 30, Sigma Aldrich), structure directing agent (SDA) 1-adamantanamine (Sigma Aldrich), ethylene diamine (Merck, Mumbai, India), LiOH flakes (Merck, India), phosphoric acid (Qualigens fine chemicals, India), morpholine (S. D. fine chemicals, India), and deionized water.

2.2 Methods

Three unlike zeolites were synthesized by three different techniques like sonochemical, sonochemical-assisted hydrothermal method, and simple hydrothermal route. In the case of synthesis of the DDR zeolite, the sonochemical synthesis approach was implemented which is assisted by the complete growth of DDR crystal in a shorter crystallization time. The precursor solution containing the molar ratio of 1 silica:0.5 1-adamantanamine:4 ethylene diamine:100 water was used in the synthesis of the DDR crystals. The details of step-by-step synthesis process for DDR zeolite was already described in the previously reported work [40]. Two different mixtures were prepared. It is reported that first the measured amount of Ludox and water were mixed together (mixture-1). Then in another mixture (mixture-2), the ethylene diamine and water were mixed in a beaker followed by addition of 1-adamantanamine. Then the final mother sol (mixture 1 + mixture 2) was sonicated for 1 h. For fast synthesis of DDR zeolite, the ultrasound equipment (UIP1500 hd HIELSCHER Ultrasound Technology) which produces acoustic waves at frequency of 20 kHz was very useful [28]. The energy input for sonication was 250 W, and the mother sol was kept for aging for 1–9 days after sonication. The powdered products were recovered through centrifugation, washed with DI water until pH < 8, and then dried in the oven at 100°C for further characterization.

In the case of Bikitaite zeolite, the molar composition of the sol used for the synthesis was 10 Li_2O:0.5 Al_2O_3:2.5 SiO_2:600 H_2O [56]. Like the previous protocol, two reactant mixtures were prepared respectively by suspending the measured amount of colloidal silica and lithium hydroxide in deionized water (DI water) in a glass beaker (mixture 1). Mixture 2 was prepared by adding the measured amount of boehmite in lithium hydroxide. Then it was mixed slowly to mixture 1 with constant and vigorous stirring, and the mixture turned into a milky white sol. The resulting mixture was sonicated for 3 h. The energy input of sonication was varied from 150 to 250 W, followed by aging for 72 h. Then the sonicated mixture was poured into Teflon-lined stainless steel autoclave. Hydrothermal crystallization was continued under autogenous pressure in a hot air oven at 100°C for 24 h. For comparison, the different Bikitaite samples were synthesized by hydrothermal process similar to the abovementioned condition without sonication treatment. After synthesis, the zeolite powders were washed thoroughly with deionized water until the pH of the washing liquid became neutral and then dried at room temperature for further characterization.

The molar composition of the sol used for the SAPO 34 zeolite synthesis was Al_2O_3:SiO_2:P_2O_5:H_2O 1:0.3:1:66. In a typical synthesis, first boehmite powder, phosphoric acid, and the required amount of water were mixed properly by using the stirrer with 600 rpm. The mixture was stirred overnight (mixture 1).

Another mixture was prepared by dissolving the calculated amount of silica sol, morpholine, and deionized water (mixture 2). Then the reaction mixture was added slowly with mixture 1, and the resulted mother solution was stirred for another 1 hour at room temperature. The resulting mixture was stirred vigorously for 15–30 min and was kept stirring overnight to produce a homogeneous sol. The prepared homogeneous sol was kept in an autoclave, and the reaction was started at 170°C for 120 h. Finally the zeolite powders were centrifuged at 12,000 rpm for 20 min followed by washing with distilled water and the same washing process repeated four times. The resultant precipitate was dried in the oven at 100°C for 1 h.

2.3 Membrane synthesis

An indigenous clay-Al_2O_3 tube of diameter 10 mm, thickness 3 mm, and 60 mm length was used as support for synthesis of the membrane. The membranes were synthesized by secondary growth hydrothermal techniques. In this technique, first the seed layer was applied on the support by using different intermediate layer in order to attach the seed crystal and prepare a uniform seed layer on the support. Then the membranes were synthesized by secondary growth of the seed layer by hydrothermal process. The membrane synthesis procedure for SAPO 34, DDR, and Bikitaite zeolites was discussed in details in our previous work [47, 50, 56].

3. Characterization of zeolite powders and membranes

The crystalline structure of the as-synthesized zeolites and membranes was determined by XRD patterns. XRD was carried out on a Philips 1710 diffractometer using CuKα radiation (α =1.541° A). The characteristic vibration bands for zeolite powders were investigated by FTIR (Nicolet 5PC, Nicolet analytical instrument, Madison, WI). Thermogravimetric analyses (TGA) and differential thermal analyses (DTA) were performed in static air using the thermogravimetric analyzer (NETZSCH STA 409 C F3 Jupiter, Germany). The samples were heated at a rate of 10°C min^{-1} under air flow. The N_2 adsorption/desorption measurements of different zeolite powder were evaluated on a volumetric gas adsorption analyzer (autosorb-iQ-MP, Quantachrome) at 77 K. The sample used in the adsorption measurement was degassed at 423 K for 6 h before the measurements. Pore size distributions and surface area data of the synthesized powders were collected from N_2 adsorption at 77 K. The same apparatus was also used for the measurement of H_2 adsorption/desorption isotherms at 77 K up to 1 bar. Prior to adsorption study, the sample was out-gassed appropriately at 250°C for 24 h under high vacuum (106 mbar). In this case, He (99.999%) and N_2 (99.999%) were used as carrier gas. Accessible microporous volume has been estimated by using the Dubinin-Radushkevich (DR) method. Transmission electron microscopy (TEM) measurements were carried out with a Tecnai G2 30ST (FEI) operating at 300 kV. The microstructure, elemental mapping with EDAX, and cross-sectional line scanning of the synthesized membranes were examined using field emission scanning electron microscopy (FESEM: model Leo, S430i, UK). X-ray photoelectron spectroscopy (XPS) measurements of support, chemically modified support, and respective membrane were carried out on an XPS system (PHI 5000 VersaProbe II, ULVAC-PHI, INC., USA) using a monochromatic Al Kα X-ray source (1486.6 eV). To identify the bonding between seed crystal and support surface, Raman analysis was carried by Raman microscope (RENISHAW inVia, UK).

The gas permeation experiment was done by a specially designed permeation cell where the membrane was mounted in a stainless steel permeation cell and sealed by silicone O-rings. Prior to permeation experiment, the leak test was carried out in order to obtain the correct data. The complete description of gas permeation measurement is given in the supporting information of our published paper [40].

4. Results and discussion

4.1 Formation mechanism of DDR zeolite

Siliceous deca-dodecasil 3R (DDR) zeolite has elliptical pore openings defined by 8- member ring windows with an effective size of 0.36 × 0.44 nm, and it is useful for separation of smaller-sized gas. DDR zeolites were synthesized by sonochemical method without the application of hydrothermal treatment. It is prepared only under sonication energy at different aging time ranging from 2 to 5 days. **Figure 2** shows all the characterization results of DDR zeolite obtained from XRD, IR, and FESEM. The XRD results of the sample explained the crystalline pattern, and the characteristic peaks are calculated by their (hkl) values. The acquired XRD patterns of the sample are most similar to that of the DDR structure, and the d-values are in agreement with those reported literature data [27]. The intensity and peak positions are well matched with the reported XRD patterns which explained the crystalline nature of the nanosized DDR zeolite. In the XRD pattern of sample aging for 2 and 3 days as shown in **Figure 2(a)**, the intensity increased gradually, and it confirms the more crystalline nature of the synthesized DDR zeolite (aging sample for 3 days).

Figure 2(b) shows the IR result of DDR crystals, and the strong vibration was noticed at 1377, 883, 767, 647, and 437 cm^{-1}. The characteristic band at 437 and 767 cm^{-1} was assigned to O–T–O (T = Si) bending and Si–O tetrahedral vibration, respectively. Here, more importantly the appearance of the peaks at 647 cm^{-1} was attributed to the double ring external linkage. The peaks at about 2915 and 2860 cm^{-1} correspond to the stretching vibration of 1-adamantanamine [44]. The symmetric stretching vibration of internal tetrahedron was shown at 747 cm^{-1}. **Figure 2(c)** shows the FESEM image of DDR zeolite, and elemental analysis showed that the desired atomic ratio of the DDR zeolite was obtained after 5 days of synthesis. The FESEM micrograph showed that the synthesized DDR seeds are nanosized powder having size 20 nm. The surface area of the synthesized powder was 212 $m^2 g^{-1}$. In sonochemical reaction process, free radicals are formed due to evolution of huge energy during collapsing of bubbles. It activates the reaction species which assisted in the nucleation and growth of colloidal nanoparticles of the reaction products [45]. In conventional synthesis process, the time required for complete reaction process is more which is often several days. But in sonochemical process, it needs less time for complete reaction process. From the XRD results, it can be assumed that the effect of

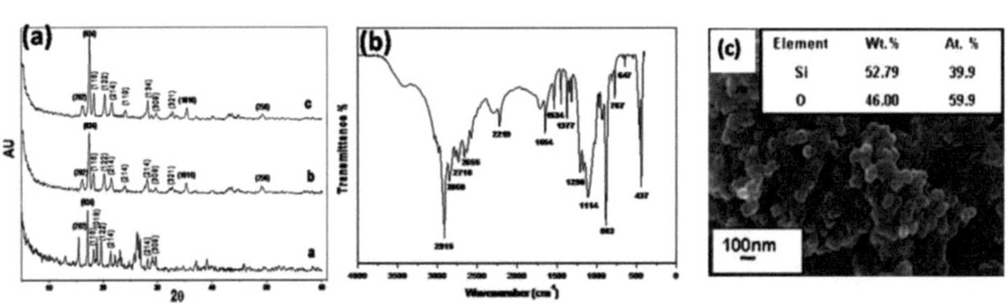

Figure 2.
(a) XRD patterns of DDR seed crystals synthesized for 2 days (lower), 3 days (middle), and 5 days (top); (b) IR spectrum of DDR zeolite synthesized for 5 days; and (c) corresponding FESEM image [43, 50].

sonication reduced the reaction time and formed crystalline DDR zeolite. The shortened reaction time is attributed to an extremely high temperature at the interface between a collapsing bubbles and the bulk solution [45]. The mass transport as well as hydrolysis and condensation reaction which were responsible for zeolite formation is controlled in this whole process. For detail, reaction mechanism and explanation in favor of this result were discussed elaborately in our reported work [42].

To develop a continuous zeolite membrane on the support surface, at first the coverage of the seed particles must be high. Generally, polycrystalline zeolite membrane contains defects (non-zeolitic pores) which are larger than zeolitic pores. The formation of non-zeolitic pores resulted from cracks and defects of the membrane layer. The non-zeolitic pores are mainly responsible for decreasing the selectivity. So in order to form a better membrane, initially seed layer plays a major role, and the adherence with the support layer is pretty much important. The lack of proper adherence of seed crystals with support may initiate the crack formation in the membrane layer. For this, different types of polymeric coating layer were used as intermediate linker between seed layer and support. The reason behind the selection of different intermediate layer and the detail mechanism were described in the literature [40, 46, 47, 50]. In the case of DDR zeolite membrane synthesis, the polymer polydiallyldimethylammonium chloride (PDADMAC) was used as intermediate linker.

PDADMAC is a high charge density homopolymer, and it can interact with several solid materials having negative surface charges. As an effect, the PDADMAC adsorbs on the alumina support primarily via electrostatic attraction between the negatively charged clay-Al_2O_3 support and the positively charged PDADMAC. Also the PDADMAC polymer can bind the negatively charged DDR zeolite particles by electrostatic attraction. As per the binding mechanism, it is assumed that due to electrostatic interaction, the negatively charged DDR zeolite particles were formed homogeneously and easily deposited on the modified support surface. This may facilitate the formation of a uniform and dense zeolite DDR membrane on the support surface. **Figure 3** shows the schematic outlook of the binding mechanism of zeolite seed layer on the support surface via using PDADMAC as intermediate linker.

The surface morphology and cross-sectional view of the synthesized DDR zeolite membrane are shown in **Figure 4(a)** and **(b)**. From the images, it can be assumed that a well-crystalline highly interlocked membrane layer was formed on the support having a uniform thickness of about 20–25 μm. Also, the phase purity of the DDR zeolite was confirmed from EDAX analysis (inset of **Figure 4(a)**), and it shows the atomic ratio of the Si and O is 1:2 which is desirable for DDR formation. The corresponding line scanning view (**Figure 4(c)** and **(d)**) throughout the membrane layer and support surface explained that the uniform membrane layer is formed on the support surface only and there is no penetration of zeolite layer into the support surface. For this outcome, intermediate PDADMAC plays a vital role. Regarding the development of continuous membrane formed or not on the support

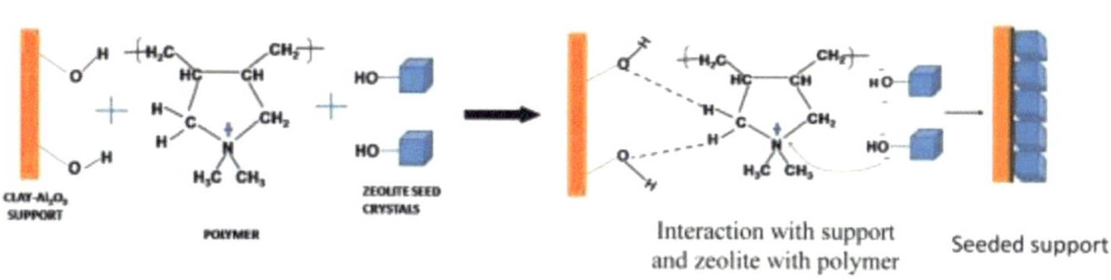

Figure 3.
Schematic of the binding mechanism between zeolite seed crystals and the support surface via PDADMAC polymer as an intermediate linker [50].

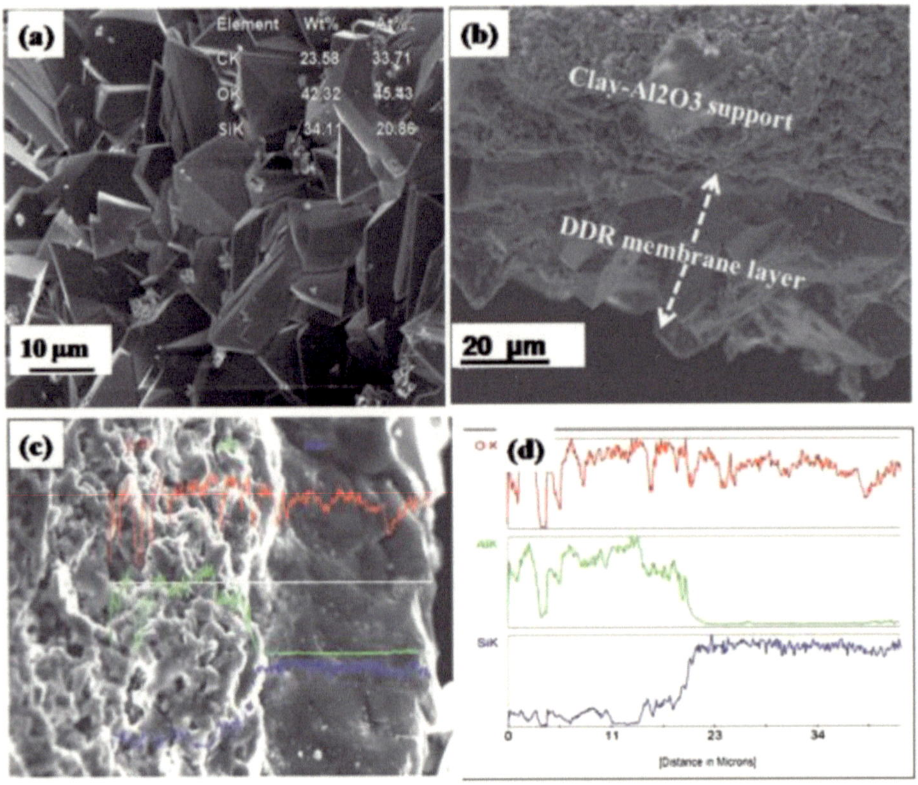

Figure 4.
FESEM images of (a) DDR membrane on the PDADMAC-modified support with EDAX data (as inset), (b) cross-sectional view of DDR membrane layer indicating the thickness by the arrow, (c) line scanning through the support and membrane layer with compositional element scan, and (d) the corresponding spectra of O, Al, and Si (distance in micron) [50].

was indicated by FESEM and XRD results. But the actual quality of the zeolite membrane can only be evaluated by gas permeation properties of the membrane.

4.2 Formation mechanism of SAPO 34 zeolite

In the case of SAPO 34 zeolite membrane, a number of studies and effort have been carried out to develop a high-quality defect-free membrane. In the literature a lot of work has been done on SAPO 34 zeolite for gas separation and storage application [32–34]. In our work an effort has been given to improvise the membrane structure and minimize the defects by different techniques for targeting the higher separation efficiency. SAPO 34 is a small pore zeolite (pore diameter of 0.38 nm), and it has a chabazite (CHA) type framework which is a promising member of the zeolite family. This interesting and efficient material has been studied for various applications. But notably its performance towards hydrogen separation from other light gases like CO_2, N_2, CH_4, etc. is quite well known [32, 33]. In the first step and prior to membrane fabrication, SAPO 34 zeolite seed crystals were synthesized by hydrothermal technique, and characterization was done systematically to know the growth of SAPO 34 zeolite during hydrothermal process. The detailed studies on the gradual formation of SAPO 34 seed crystals at different times ranging from the initial gel mixture to 120 h at 170°C were described in our previous work [51]. The XRD pattern of the zeolite powders synthesized for 120 h is shown in **Figure 5(a)**. The complete crystallization of SAPO 34 was noticed only after 120 h of hydrothermal synthesis at 170°C. All diffraction peaks are mostly similar to those of the chabazite structure of SAPO 34, and the d-values, i.e., (100), (110), (210), (220), (211), and (131), are in agreement with those reported in the literature [32]. **Figure 5(b)** shows the FESEM image of SAPO 34 seed crystals and almost all the crystals are cubical in nature having size 2-3 micron.

The TEM image of SAPO 34 powder after 120 h of hydrothermal synthesis is shown in **Figure 5(c)**, which looks like a cubic structure, and corresponding SAED pattern reflected the CHA structure of SAPO 34 zeolite as shown in **Figure 5(d)**. **Figure 5(e)** describes the FTIR spectra of the SAPO 34 zeolite powders collected at different time starting from 0 (initial gel) to 120 h, during seed synthesis by hydrothermal method. The various zeolitic vibration frequencies were assigned in accordance to the reported literature [1]. From this study, it was confirmed that after 120 h, the complete chabazite structure of SAPO 34 zeolite is formed. The characteristic band at 480, 534, and 568 cm^{-1} was attributed to the vibration of SiO_4, (Si, Al) O_4, and PO_4, respectively. In addition, the vibration peak at 638 cm^{-1} matched with the double-6 rings (D6) would be the key evidence to CHA framework completeness. The gradual formation of SAPO 34 results at different time period, and the detail characterization was explained properly in the literature [51]. In the case of membrane synthesis, SAPO 34 zeolite membranes were synthesized on the clay-Al_2O_3 support by ex situ (secondary growth) hydrothermal method. In our approach we have prepared SAPO 34 zeolite membrane which is composed of three parts: the substrate, the intermediate layer, and the seed layer.

The intermediate layer is composed of the polymeric/inorganic oxide layer with dispersed zeolite seed crystals. The importance of intermediate layer used

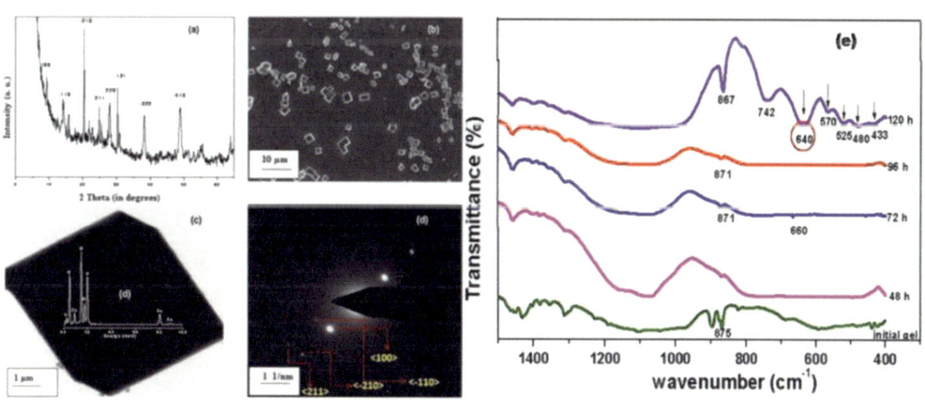

Figure 5.
(a) XRD pattern of SAPO 34 zeolite synthesized by hydrothermal technique at 170°C for 120 h, (b) FESEM image of SAPO 34 zeolite, (c) bright field TEM image synthesized powder and inset show the EDS results, (d) corresponding SAED pattern of SAPO 34 zeolite, and (e) FTIR spectra of SAPO 34 zeolite samples synthesized at different time, ranging from 0 (initial gel mixture) to 120 h of hydrothermal synthesis [46, 51].

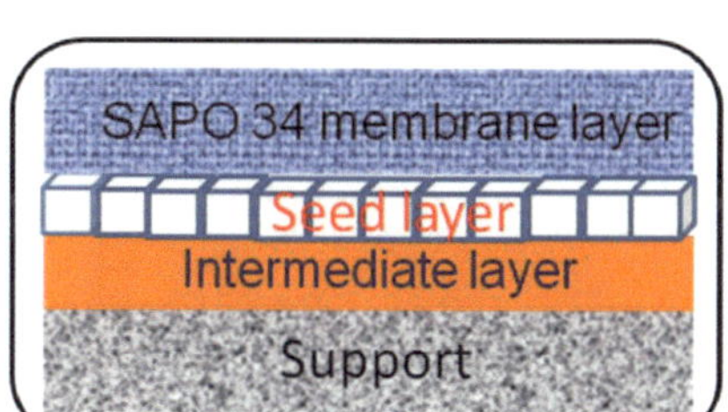

Figure 6.
Schematic representation of the different steps involved in SAPO 34 membrane syntheses.

for membrane synthesis was discussed in our reported results [40, 46, 47, 50]. The schematic representation for SAPO 34 membrane preparation is given here for better understanding (**Figure 6**).

In SAPO 34 membrane synthesis, silica intermediate layer was proposed for the selective deposition of oriented zeolite seed crystals with closely packed monolayers on a low-cost clay-Al_2O_3 tubular support and then subjected to the secondary growth under suitable hydrothermal conditions. As a result, a homogenous and reduced defect in highly oriented membrane can be synthesized simply. The organization of zeolite microcrystals with controlled orientation on substrates has been a subject of scientific interest, and recently, several approaches have been developed to prepare zeolite films with controlled orientation [52–55].

Significantly, using silica intermediate layer is predominantly based on the thermal and mechanical stability, as well as it being able to withstand in very high pressures. The silica surface under normal conditions is recovered with reactive hydroxyl groups, Si–OH, called silanol groups, and the high density of such surface –OH groups may promote high coverage of the resulting zeolite film and the highly oriented crystals. Furthermore, the implication of a silica layer plays an important role during membrane development. First of all, the layer helps to make the support surface smoother for the deposition of seed crystals in a uniform direction, and most importantly it acts as a blockade layer for the penetration of zeolite seed crystals into the interior of the support. Also it facilitates to persist the support layer with more hydroxyl (–OH) groups and as a result imparts the support with more nucleation points where crystals could bind on the support via van der Waals interactions and H-bonding [55]. In addition, it reduced stress-induced crack formation at the support–zeolite interface during calcination step [52]. The synthesized membranes were characterized by XRD, FESEM, and FESEM elemental mapping, etc. Finally the actual quality of the membrane was evaluated by gas permeation studies.

The formation of phase pure SAPO 34 membranes with a high degree of crystallinity and correct orientation was confirmed by XRD analysis. **Figure 7(a)–(d)** shows the XRD patterns of the modified and nonmodified substrate, along with the membrane layer on that substrate. SAPO 34 membrane layer prepared on the

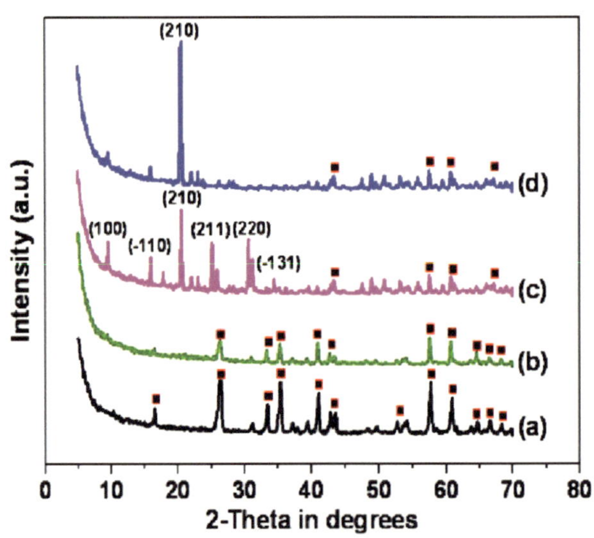

Figure 7.
XRD patterns of (a) bare support, (b) silica-modified support, (c) SAPO 34 membrane on the nonmodified support, and (d) membrane on the silica-modified support synthesized at 175°C for 120 h by hydrothermal process. (▪) peak from the clay-Al_2O_3 support [40].

nonmodified support shows that all diffraction peaks are most similar to those of the chabazite structure of SAPO 34, and the intensity of the peaks like (220), (211), (131), etc. are very much prominent which indicates the presence of pure SAPO 34 crystals from the membrane surface with random orientation.

In the case of XRD pattern of the SAPO 34 membrane prepared on the silica-modified support, a quite interesting result was perceived, that is, the intensity of the (210) peak which increases drastically compared with other peaks. It can be explained that the membrane developed towards higher orientation. In addition, the membrane layer also associates with small amount of non-oriented crystals [53]. The stronger intensity of the single peak proves the possibility of formation of an oriented membrane on the silica-modified support surface which plays a vital role in determining the high performance of the membrane. The same interpretation was noticed from the FESEM results. **Figure 8(a)** and **(b)** illustrates the FESEM micrograph of the bare support and silica-modified clay-Al_2O_3 support surface, respectively. The homogeneous oriented seed monolayer on the silica-modified support is shown in **Figure 8(c)**. From the FESEM images, it can be explained that the majority of the seed crystals are deposited with proper orientation along with some disoriented seed particles interfering during the seeding process. Then during hydrothermal synthesis at 170°C for 120 h, the seed layer grows epitaxially and formed an oriented membrane layer on the support.

Figure 8(d) depicts the surface morphology of the SAPO 34 membrane layers. It shows that the support surface was totally covered by uniform and compact

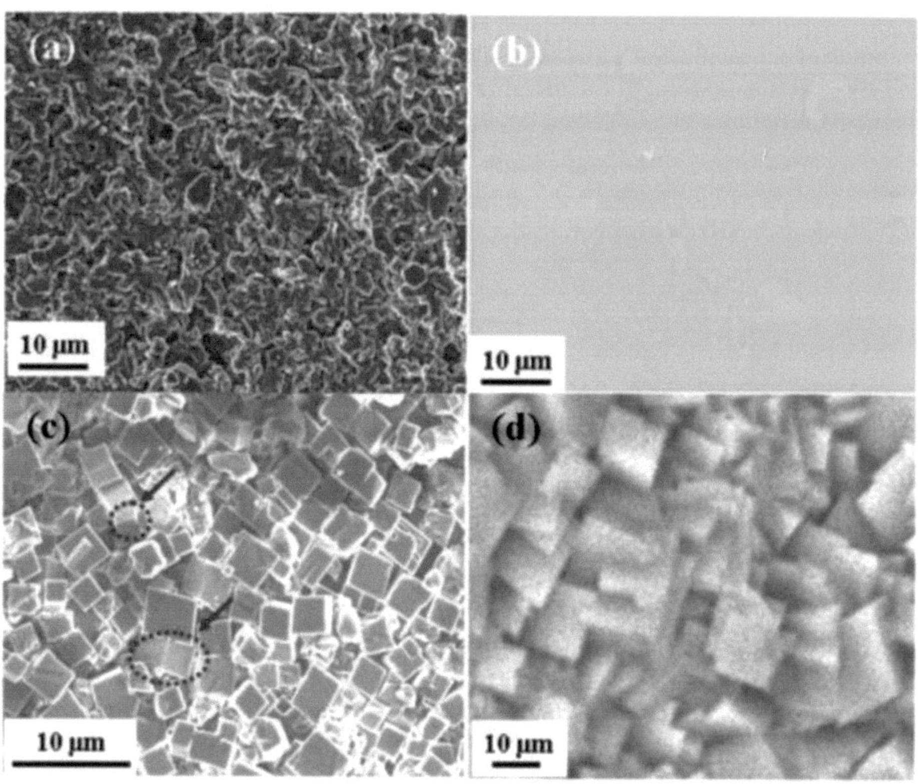

Figure 8.
FESEM micrographs of (a) clay-Al_2O_3 substrate, (b) top view image of silica-modified support, (c) oriented seed monolayer on silica-modified support with a few misoriented seeds indicated by arrow marks, and (d) oriented SAPO 34 membrane layer synthesized on the seeded support prepared at 175°C for 120 h by a hydrothermal process [40].

cubic-shaped crystals, and no visible cracks, pinholes, or other macroscopic defects were observed. The FESEM micrograph of SAPO 34 membrane layer prepared on the nonmodified support surface has already been described earlier. The comprehensible discussion was described in our published paper [46]. This work is an attempt and is anticipated to have much importance for making defect-free low-cost highly oriented membranes which could offer to be a safe, simple, and environmentally benign potential application for gas separation application. Then ultimate gas separation performance of the SAPO 34 membrane is discussed to explain the membrane efficiency.

Also another successful approach was implemented to develop an oriented defect-free membrane on the support surface. The oriented SAPO 34 membranes were grown on the support using a secondary (seeded) growth hydrothermal technique followed by insertion of 11-mercaptoundecanoic acid capped palladium (MUA-Pd) nanoparticles to the membrane.

Prior to membrane synthesis, first the clay-Al_2O_3 support was treated with polydiallyldimethylammonium chloride (PolyDADMAC) polymer, followed by deposition of seed layer homogeneously in a regular orientation on the support surface. A schematic representation of the membrane synthesis processes starting from bare support to nanoparticle insertion is shown in **Figure 9**. To deposit Pd NPs in the membrane matrix, a simple dip-coating technique was used. In practical, during thermal treatment of the Pd/SAPO 34 membrane, the instigation of defects is because of the removal of structure directing agent (SDA) from the zeolite pores. But interestingly, the presence of Pd NPs which entrapped inside the non-zeolitic pores and clogged the defects of the membrane. The synthesized membranes were characterized by XRD, TEM, XPS, and FESEM technique. FESEM and elemental mapping of the membrane cross section confirmed that most of the Pd NPs were deposited at the interface of the membrane and the support layer which may increase the membrane efficiency, i.e., separation efficiency.

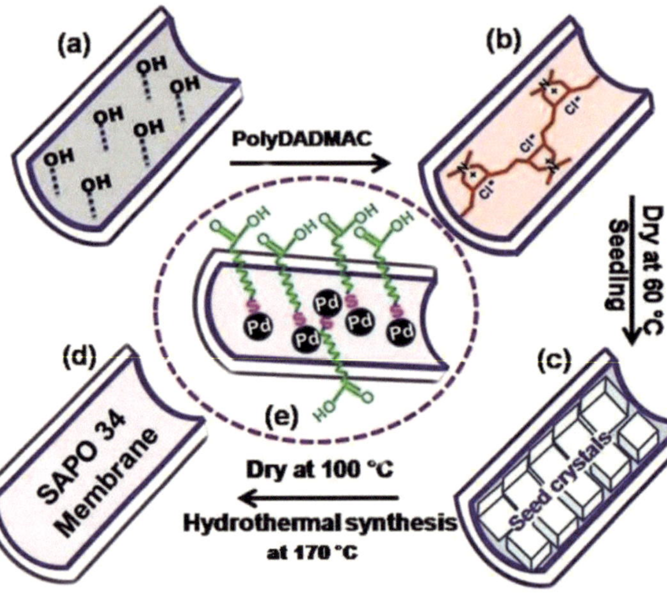

Figure 9.
Schematic representation of growth of SAPO 34 membrane starting from (a) bare substrate, (b) PolyDADMAC-modified layer to capture zeolite seed crystal, (c) seed monolayer onto the modified support, (d) synthesized membrane by secondary growth hydrothermal process, and (e) membrane layer decorated with MUA-capped Pd NPs [47].

The formation of phase pure, highly crystalline SAPO 34 zeolite on the support surface and the existence of Pd NPs on the membrane surface were confirmed by X-ray diffraction (XRD) patterns as revealed in **Figure 10**. From the XRD pattern of Pd/SAPO 34 membrane, it is clear that the intensity of the (210) peak is higher than other (100), (−110), (220), (211), and (−131) peaks and proves the presence of oriented crystals in the membrane layer. The Pd face-centered cubic phase has been identified from the Pd (111) peak which confirms the presence of Pd NPs in the membrane layer. The interpretation from the XRD results explained the presence of the Pd NPs, and the vital information is that the presence of Pd NPs did not affect the crystal structure of the SAPO 34 zeolite. It is because of the implicated nanoparticles which are too large to reside in the cavities (0.38 nm) of the framework. In general, during the heat treatment process, the structure directing agents (SDA) or any other organics are removed, and as a result, non-zeolitic pores, i.e., intercrystalline gaps, defects, or cracks, are formed.

Hence, it may be believed that the non-zeolitic pores were occupied by Pd NPs during the thermal treatment of the membrane, and further interpretation was established by FESEM studies, EDAX analysis, and elemental mapping. These results were explained in our earlier work [47]. For clear understanding, FESEM results were described here. **Figure 11(a)** depicts a FESEM micrograph of the Pd/SAPO 34 membrane layer prepared on the PolyDADMAC-modified support. It appears that a uniform membrane layer was formed with an interlocking structure. The uniformity of the membrane was achieved due to the coverage and proper orientation of seed crystals on the support which ultimately facilitated the formation of high-quality membrane. The high magnification FESEM data tells that no visible cracks, pinholes, or other macroscopic defects were noticed on the membrane layer. Then to know the membrane structure after thermal treatment, further FESEM characterization was done, and the micrograph of the calcined Pd/SAPO 34 membrane is illustrated in **Figure 11(b)**. The membrane surface was analyzed carefully by selecting different areas.

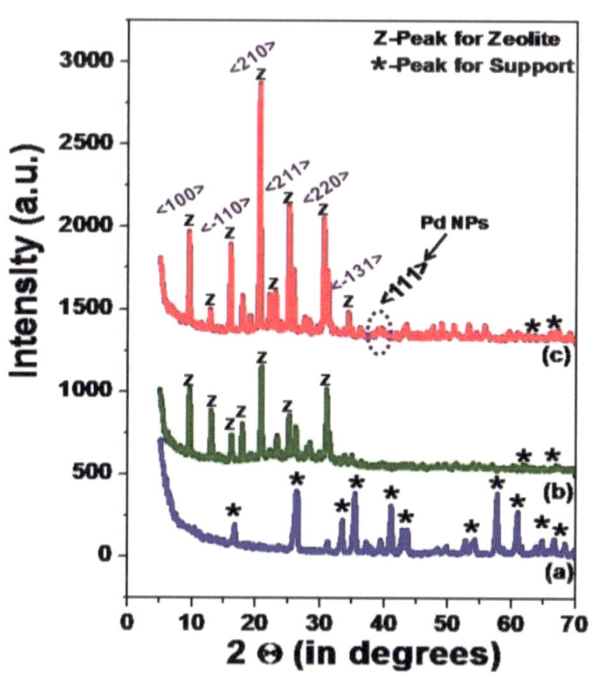

Figure 10.
XRD patterns of (a) clay-Al_2O_3 support, (b) the SAPO 34 membrane layer prepared on the nonmodified support, and (c) the Pd/SAPO 34 membrane layers on the modified support synthesized by the secondary growth hydrothermal technique [47].

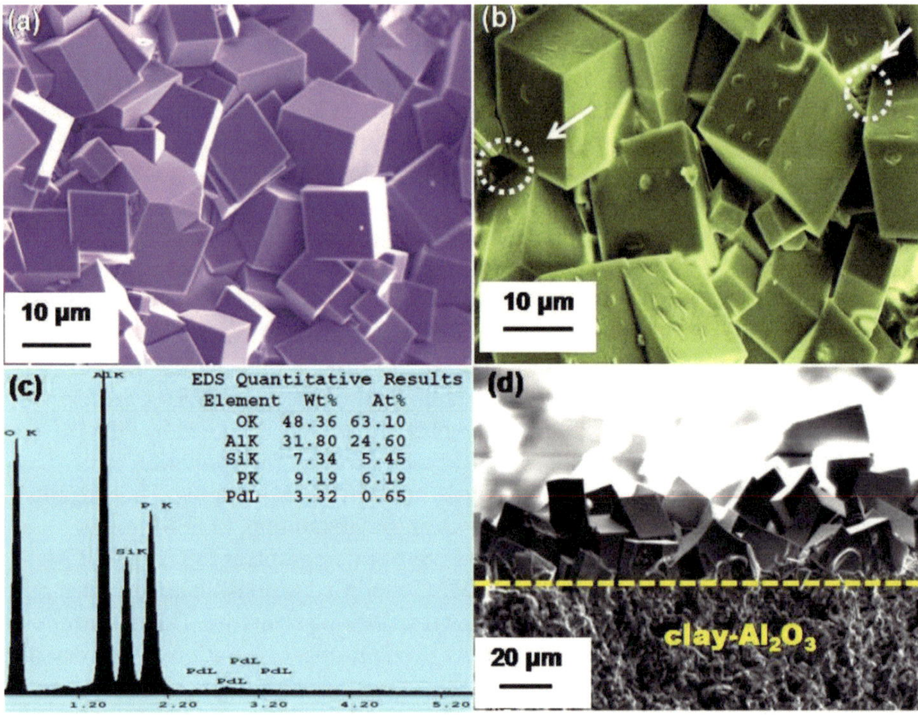

Figure 11.
FESEM micrographs of (a) Pd/SAPO 34 membrane layer, (b) calcined Pd/SAPO 34 membrane (dotted circle indicated by arrow mark shows the defects formed after calcination processes), (c) corresponding EDS spectra taken from the selected area indicated by dotted circle and the inserted table show the quantitative analysis, and (d) cross-sectional view of the Pd/SAPO 34 membrane layer [47].

During inspection, the defective areas were identified, and to verify the presence of nanoparticles inside the defective area, EDAX analysis of the same area was done (**Figure 11(c)**). The Pd peak was identified along with SAPO 34 zeolite, and Si/Al ratio confirmed the complete growth of SAPO 34 zeolite. The quantitative elemental analysis of the synthesized Pd/SAPO 34 membrane coating is described in the table inset in **Figure 11(c)**. The FESEM cross-sectional view (**Figure 11(d)**) shows that the thickness of the membrane is ~20–25 micron. The FESEM study revealed that the interlocked dense membrane was formed on the support surface successfully. However, the presence of Pd NPs which occupied and plugged into the non-zeolitic pores assisted towards the development of a nearly reduced defect membrane. It is noticed that two important phenomena were carried out simultaneously during thermal treatment. First, in the calcination process formation of non-zeolitic pores during the removal of structure directing agent, and simultaneously in the second step Pd NPs are migrating and entrapped inside the non-zeolitic pores and clogged the defects. Finally, in order to check the Pd/SAPO 34 membrane quality, gas permeation studies were carried out at room temperature in different feed pressures.

This work highlighted how the non-zeolitic pores of the synthesized membranes can be repaired by the insertion of palladium nanoparticles in the membrane matrix. Orientation and drastic reduction of non-zeolitic pores in the membrane layer may enhance the membrane quality for gas separation application.

4.3 Formation mechanism of BIK zeolite and modified BIK zeolite

Historically, activated carbons and zeolites have been the most studied microporous materials (pore diameter < 2 nm) for the storage of gases. Zeolites were the first materials looked to as adsorbents for ANG technologies and methane adsorption in zeolites continue to assist in the understanding and design of adsorbent materials [57–58]. Zeolites as hydrogen storage materials are investigated broadly, and it is found that small molecules such as molecular hydrogen can be easily absorbed into a flexible network of zeolites, and hydrogen adsorption properties of different types of zeolites have been investigated [59–63]. It was found that the amount of hydrogen adsorbed on zeolites can be affected by the framework composition, structure, charge-compensating cations, and acidic–basic nature. In order to create strong binding sites for H_2 in zeolite pores, the importance of charge balance is quite inevitable. In this perspective, the role of light alkali metal cations such as Li^+, Na^+, and Mg^{2+} into the porous framework of zeolite plays an important role and enhances the binding energy for hydrogen adsorption [64]. Among the alkali metal cations (Li^+, Na^+, Mg^{2+}), lithium ion is more capable due to its low atomic weight and high affinity towards hydrogen by charge-induced dipole interactions [65]. Based on this concept, Li substituted Bikitaite zeolite nanocrystals are synthesized at room temperature in short crystallization time by sonochemical method. Further, ultrasonic irradiations have been used along with hydrothermal treatment for synthesis of zeolite nanocrystals. First the synthesized powders were characterized by XRD and confirmed the gradual formation of highly crystalline material. **Figure 12** shows the XRD pattern of sonochemically synthesized Bikitaite zeolite at different sonication time starting from 1.5, 2, and 3 h, and the sonication energy of 150 W was fixed for the synthesis. Before sonication, the sol was aged for 72 h at room temperature. The XRD result reveals that after 1.5 h sonication, the zeolite phase started forming and remained the same up to 3 h of irradiation. But complete growth of nanocrystalline Bikitaite phase was observed after hydrothermal treatment of sonicated sol for 24 h at 100°C. The figure clearly shows that the nanocrystalline Bikitaite zeolite with major peaks (100), (101), (201), etc. has formed and all the XRD patterns were compared with XRD pattern of simulated zeolite (COD file-969,003,103) as described in the literature [56]. Next, the morphology of the powders was observed by FESEM. **Figure 12(b)–(e)** depicts the morphology and corresponding EDAX analysis of the powders synthesized by only sonochemical method. Comparing these micrographs, it is clear that, at room temperature by ultrasonic irradiation, Li zeolite was formed with smaller size. But after further hydrothermal treatment,

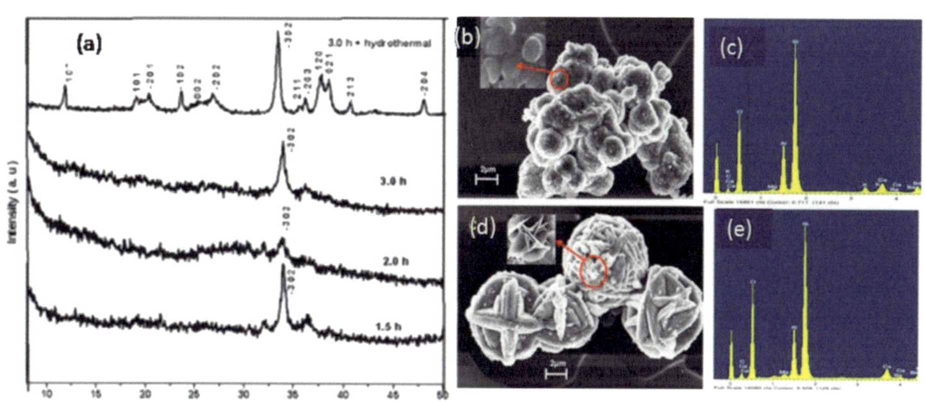

Figure 12.
(a) XRD pattern of Bikitaite zeolite synthesized with ultrasonic irradiation at different time, sonication energy of 150 W, aging time 72 h and sonication followed by hydrothermal treatment for 100°C and 24 h, FESEM micrograph of Bikitaite, prepared by (b) sonication for 3 h (inset: Higher magnified picture), (c) corresponding EDAX, (d) sonication followed by hydrothermal treatment at 100°C for 24 h (higher magnified picture is inset), and (e) corresponding EDAX [56].

primary particles formed, reflecting porous woolen ball-like structures, which finally formed flake-like flower structure (**Figure 12(d)** inset).

The EDAX result (**Figure 12(c)** and **(e)**) explained that both phases show that the presence of silica is more than alumina which is in accordance to the reported stoichiometry of Bikitaite [66].

4.4 Gas storage

Hydrogen adsorption capacity of the developed materials suited at cryogenic temperature and room temperature. The highest H_2 adsorption capacity for pure Li zeolite reached up to 1.3 wt% which is more than the reported value. The synthesized zeolite was characterized by different techniques, and then the adsorption study carried out using appropriate method. The lithium doped sample showed higher hydrogen sorption capacity showed nearly 1.3 wt% compared with other zeolite as described in literature at 77 K and 1 bar pressure. **Table 1** shows comparison of H_2 adsorption capacity of Bikitaite with other reported values of zeolite [56]. The detailed study and explanation for hydrogen storage in Bikitaite zeolite have been discussed in our recent publication [67, 68].

Finally, the study has shown that Bikitaite zeolite is a promising material for hydrogen storage. The storage volume increases with increasing Li content of the zeolite. This can be attributed by strong interaction between hydrogen molecule and high charge density of Li + ion. The detailed description of this work was reported earlier [56]. For membrane fabrication, the same powders were used as a seed for secondary growth. The mechanism and detailed procedure of membrane fabrication and their characterization results were discussed properly in the literature [69]. In this book chapter, only SAPO 34 and DDR membrane are highlighted.

Zeolite	Temperature (K)	Pressure (bar)	H_2 adsorption capacity	
			c.c/g	(wt%)
H-SAPO-34	77	0.92	122.08	1.09
H-Chabazite			123.2	1.10
(Ca-Na) Y	77.3	1.01	100	0.89
Ca-X	77.3	1.01	130	1.16
NaA	77	1.01	76.16	0.68
KA	77	1.01	11.2	0.1
CaA	77	1.01	67.2	0.6
NaX	75	1	94.08	0.84
CaA			108.64	0.97
NaA			88.48	0.79
NaX	100	60	23.52	0.21
FAU	77	1	100.8	0.9
ZSM5	77	1	80	0.71
Zeolite-L			59	0.53
MCM-41			65	0.58
Ferrierite			65	0.58
BIK	**77**	**1.01**	**143.2**	**1.3**
BIK	**298**	**1.01**	**1.0**	**0.009**

Table 1.
Comparison of H2 adsorption capacity of Bikitaite zeolite with reported values.

4.5 Gas separation

In this part gas permeation and separation studies of synthesized DDR and SAPO 34 membrane were discussed. In order to upgrade the quality of SAPO 34 zeolite membrane, different approaches have been taken as described previously. Successfully "reduced defect" or nearly "defect-free" oriented membranes were synthesized on the low-cost clay-Al_2O_3 support. The separation performances of the synthesized membrane designate whether a high-quality membrane was formed or not. Therefore the synthesized membranes were used for gas separation studies at different feed pressures as well as different feed compositions. The separation studies were carried out at room temperature.

In the case of DDR membrane, the hydrogen separation efficiency was evaluated. Before gas permeation studies, the membranes were calcined to remove all the structure directing agents and organic compounds present in the zeolitic pore.

The permeance is expressed as the flux rate through all the pores present in the membrane. The diffusion rate becomes significantly smaller where the kinetic diameter of the gas becomes larger than the pore size of the zeolite. The molecular kinetic diameters of H_2 and CO_2 are 0.29 and 0.33 nm, respectively, which are close to the pore size of DDR zeolite. The configurational diffusion and the variation in molecular size between H_2 and CO_2 result in the difference in the rate of diffusion through the DDR zeolite channels. The diffusion rate of H_2 is faster than that of CO_2, and therefore, H_2 and CO_2 can be separated by DDR zeolite membrane. **Figure 13** describes the change of permeated flux of both H_2 and CO_2 through DDR zeolite membrane with varying transmembrane pressure difference. It shows that the rate of change of permeating flux with pressure is less for both CO_2 and H_2. In the case of molecular sieving through zeolitic pores, the rate of change of flux is pressure independent. But in this result, the little increase of flux with pressure shows that the membrane is associated with low concentration of non-zeolitic pores. At high pressure, CO_2 adsorbs more strongly than H_2 on the DDR zeolite membrane surface due to quadruple moment nature of CO_2. So rate of desorption of CO_2 from the membrane surface also decreases; as a result, permeating flux also decreases compared with hydrogen.

The real performance of the membranes can be enlightened by their mixture gas separation ability. **Figure 14** assigns the H_2/CO_2 separation factor of the mixture gas at room temperature as a function of CO_2 feed concentration at 200 kPa feed pressure for the DDR membrane. From mixture gas, it is explained that selectivity decreases with increasing CO_2 concentration. In general, CO_2 has been preferentially adsorbed on the DDR pore surface, and therefore with increasing CO_2 concentration, the extent of pore coverage has also increased. As a result, H_2 permeability decreased and selectivity reduced spontaneously. The competitive adsorption-diffusion mechanism along with molecular sieving both is playing an important role for separation process. The H_2/CO_2 separation selectivity of the membrane

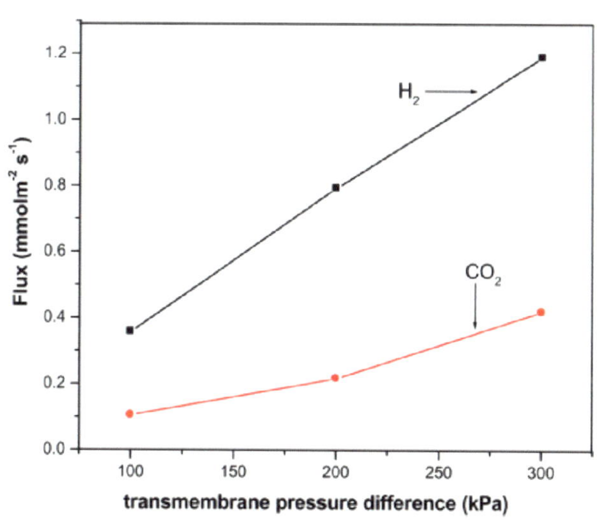

Figure 13.
Single gas permeated flux of H_2 and CO_2 as a function of transmembrane pressure difference [50].

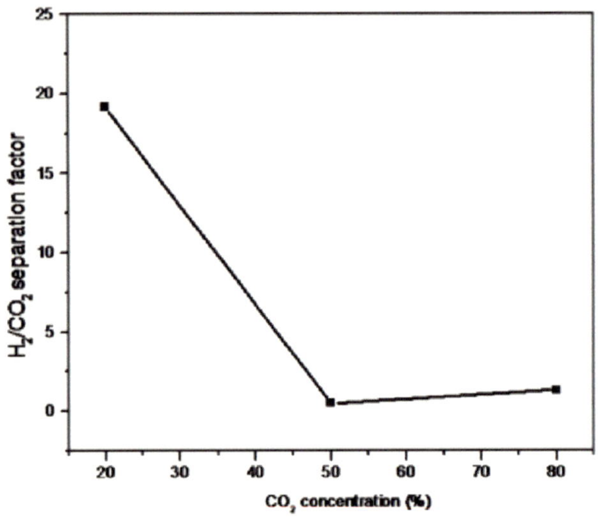

Figure 14.
H_2/CO_2 separation factor of the mixture gas at room temperature as a function of CO_2 concentration in feed at 200 kPa feed pressure for DDR membrane [50].

increased up to 3.7 at room temperature which is more than the reported values and separation mechanism explained properly in our reported work [46].

But some appreciable gas separation results were found in the case of SAPO 34 membrane synthesized on SiO_2 modified support. The SAPO 34 membrane synthesized on SiO_2 modified support shows an appreciable hydrogen separation from CO_2 and N_2. **Figure 15** shows the single gas permeation of H_2 and CO_2 at room temperature at different feed pressures. The synthesized membrane shows a relatively high hydrogen gas permeation value as compared with the literature values. Several factors might contribute to the higher hydrogen gas permeance through SAPO 34 zeolite membranes in this study. In general, the preferred orientation of the membrane layer plays a significant role in gas permeation. Generally,

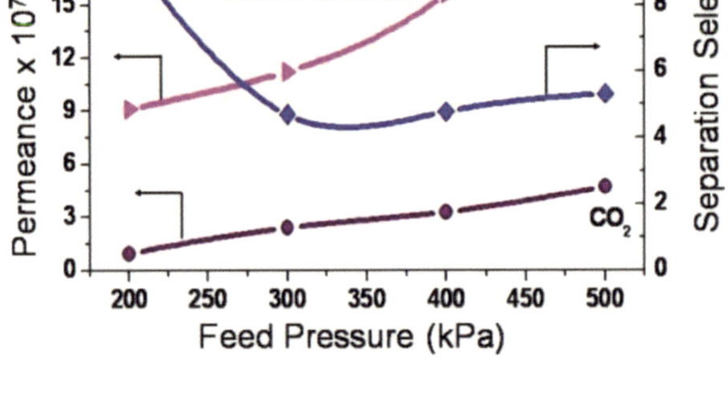

Figure 15.
Single gas permeation through SAPO 34 membrane at room temperature as a function of different feed pressures and separation selectivity of H_2/CO_2. The inset shows the comparison of separation selectivity of H_2/CO_2 by using SAPO 34 zeolite membrane prepared on the modified and nonmodified support surface [40].

oriented porous paths exhibit superior performances compared with tortuous paths of randomly oriented pores of a membrane layer by minimizing the defect density and membrane resistance.

Actually, in the case of a highly oriented membrane structure, the pores are more aligned, and the resistance of the gas transport through the aligned channel is less than that of the zigzag path of the nonaligned randomly oriented pores. As a result, membrane resistance decreases and the permeation adequacy enhanced, as compared with the randomly oriented membrane. As per our earlier discussion, the use of an intermediate silica layer which impedes the penetration of the zeolite seed particles inside the pores of the support increases the ultimate permeability of the membrane. So, combining all these aspects, it can be concluded that the almost defect-free, highly improved SAPO 34 membrane was developed on SiO_2 modified support which shows higher hydrogen gas permeance with creditable results. The configurational diffusion and the difference in molecular size between H_2 and CO_2 result in a difference in the rate of diffusion through the SAPO 34 zeolite channels. The diffusion rate of H_2 is faster than that of CO_2. The H_2/CO_2 selectivity gradually increases with respect to the different feed pressures. More interestingly, at room temperature, the appreciable highest selectivity value for H_2/CO_2 was found to be 9.12. As shown in the inset of **Figure 15**, it describes the comparative study of selectivity for H_2/CO_2, and the values were lower through the SAPO 34 membrane prepared on the nonmodified support with respect to the modified support under different feed pressures. This result indicates that there are fewer non-zeolitic pores in the case of the SAPO 34 membrane on the modified support. However, the lowest selectivity values strongly indicate the presence of non-zeolitic pores, i.e., defects in the membrane layer.

The real performance of the membranes can be investigated by their gas mixture separation ability (**Figure 16**). In the case of the H_2/CO_2 system, an appreciable selectivity value of 16.66 was obtained, and selectivity gradually decreased with increasing feed pressure. This phenomenon can be explained by the same way, i.e., an adsorption-diffusion model. At higher pressure, CO_2 adsorbs more preferentially than H_2 because it has the strongest electrostatic quadrupole moment, and more adsorption sites are generated which block some of the adsorptions of more weakly absorbing species. As CO_2 adsorbs preferentially in the SAPO 34 pore wall, it also

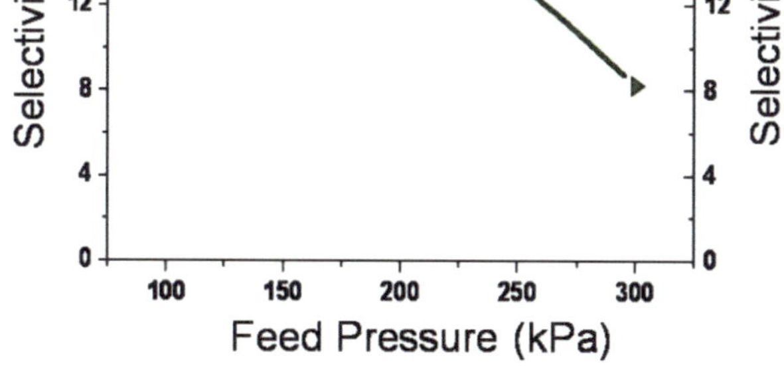

Figure 16.
Room temperature separation selectivity of 50:50 H_2/CO_2 and 50:50 H_2-N_2 mixture as a function of different feed pressures. (flow rate = 100 mL min^{-1}) [40].

desorbs and diffuses earlier than H_2, and the reasonable overall selectivity value in mixture was decreased. In H_2-N_2 gas mixture separation, a selectivity value of 20.91 was achieved and decreased to a small extent with respect to the feed pressures. Here, N_2 has a negligible effect with respect to CO_2 on H_2 during the separation process, and because of the larger size of the N_2 (0.36 nm) than CO_2 (0.36 nm), the effect of molecular sieving plays a major role in the higher selectivity. The combined effect of these two determines the ultimate selectivity of the membrane. The obtained selectivity values are improved compared with the reported literature values as described in earlier data [40]. Fascinatingly, the selectivity values were performed at high values and remain almost constant up to 80 h. The high reproducibility was due to the formation of defect-free and highly oriented membrane layers.

Next, permeation results were given here to understand the Pd/SAPO 34 membrane quality. The flow rate of different gases was controlled by the mass flow controller (MFC). **Figure 17** describes the single gas permeance of different gases through SAPO 34 membranes prepared on modified supports with and without Pd loading at 30°C and 200 kPa feed pressure as a function of the gas kinetic diameter (nm). It is interesting to note that single gas permeance through the Pd/SAPO 34 membrane at room temperature changed dramatically in comparison to the SAPO 34 membrane without Pd loading. As the kinetic diameter increased, the difference in the permeability of gases through those membranes decreased because of the difference in the kinetic diameter of gas molecule.

Mostly, the significant reduction of hydrogen permeance in the case of the Pd/SAPO 34 membrane as compared with the SAPO 34 membrane indicated the drastic reduction of non-zeolitic pores. However, in the case of the SAPO 34 membrane, the higher permeance value of H_2 was obtained because of the presence of defects. From this result, it can be concluded that the non-zeolitic pores are at a minimum in the SAPO 34 membrane after Pd NPs loading. **Figure 18** describes the change of single gas permeance of H_2 and CO_2 through the SAPO 34 and Pd/SAPO 34 zeolite membrane at different feed pressures. It can be explained that the rate of increase of the permeance values of CO_2 with feed pressure is less than that of H_2, and the explanation for this result has already been discussed. As CO_2 adsorbs preferentially on the SAPO 34 zeolite membrane surface than H_2, so the rate of desorption of CO_2 from the membrane surface also decreased as compared with H_2.

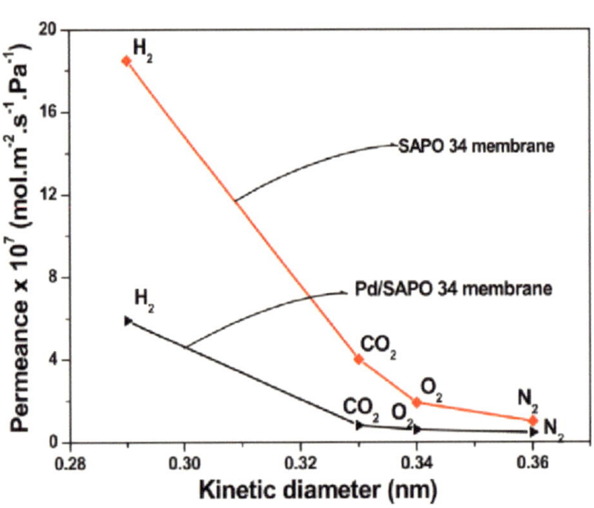

Figure 17.
Single gas permeances of different gases through SAPO 34 and Pd/SAPO 34 membranes prepared on modified supports at 30°C and 200 kPa feed pressure as a function of the gas kinetic diameter (nm) [47].

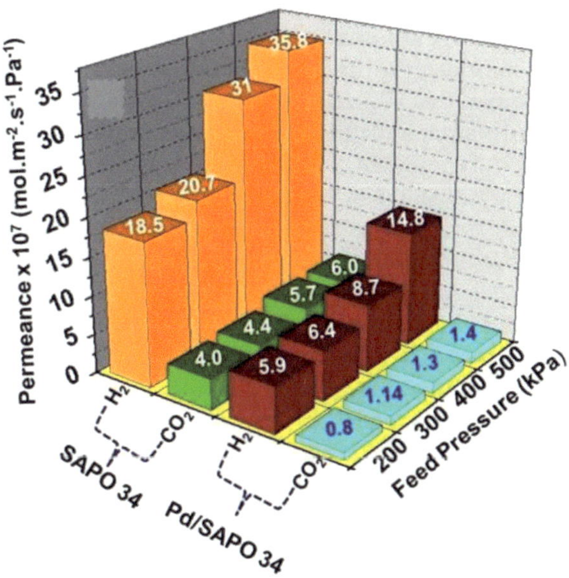

Figure 18.
Room temperature single gas permeation study of SAPO 34 and Pd/SAPO 34 membrane at different feed pressures. (flow rate 100 mL min^{-1}) [47].

Hence, due to the preferential adsorption and diffusion of CO_2 on the SAPO 34 zeolite surface and the difference in molecular size between H_2 and CO_2, there is a difference in the permeance values through the SAPO 34 zeolite channels.

According to **Figure 18**, as expected, the H_2 permeability through the Pd/SAPO 34 membrane was lower than that through the SAPO 34 zeolite membranes, and the drastic reduction of the H_2 permeance value as compared with SAPO 34 indicates that non-zeolitic pores were repaired by Pd NPs. Again, in the case of Pd/SAPO 34, the CO_2 permeance values are almost equal at different feed pressures which confirm the removal of non-zeolitic pores and permeability through the Pd/zeolite membrane mainly due to the molecular sieving process which is less dependent on feed pressure.

The real performance of the membrane for hydrogen gas separation from a mixture was evaluated from their mixture gas separation studies. The highest mixture gas separation factor for the Pd/SAPO 34 membrane was achieved at 20.8 [47]. However, the mixture separation factor in the case of the SAPO 34 membrane was 6.2. Both values, i.e., hydrogen permeation and separation factors, were higher than the literature values. Separation selectivity of the SAPO membranes increases appreciably after insertion of the Pd NPs which can reduce the non-zeolitic pores to a large extent and improve the membrane quality for hydrogen gas separation.

5. Concluding remark and outlook

We have summarized our ongoing research endeavors of microporous zeolite material for efficient gas storage and separation application. The key efforts that have been made mainly involve the synthesis of zeolite nanocrystal in a shorter crystallization time and develop the high-quality-oriented membrane on the low-cost support. To develop a high-quality membrane, different approaches have been taken and can be stated that our approach towards the development of high quality membrane is successful and may be applicable for other microporous materials for membrane development. The detail explanation and other fundamental understanding are described in our reported paper. Also many techniques are proposed by different research groups towards the development of ideal membrane and the clear understanding can be acquired from the literature. The hydrogen storage ability of Bikitaite zeolite is really commendable, and the H_2 separation performance from H_2-CO_2 and H_2-N_2 mixture for DDR and SAPO 34 zeolite membrane is appreciable as compared with reported data. However, it should be noted that there are still many issues remaining to be addressed before implementation of industrial and commercial usage of these developed materials for separation and storage purpose. For these issues, chemists and engineering scientists of different expertise and industrial partners need to work cooperatively. In realistic separation systems, the gas mixtures and operation conditions are more complicated, and many factors need to be taken into account to perform successfully under industrial relevant condition. There is no doubt that zeolite materials are an area of great excitement and potential importance in all the areas. It would be a great achievement for such materials to be applied in practice.

Acknowledgements

The authors would like to thank CSIR, India, and also thankful to Dr. K Muraleedharan, Director, CSIR-CGCRI, for his kind permission to publish the book chapter.

Author details

Nandini Das* and Jugal Kishore Das
Ceramic Membrane Division, CSIR - Central Glass and Ceramic Research Institute, Kolkata, India

*Address all correspondence to: dasnandini@cgcri.res.in

References

[1] Morris RE, Wheatley PS. Gas storage in nanoporous materials. Angewandte Chemie, International Edition. 2008;**47**:4966-4981. DOI: 10.1002/anie.200703934

[2] Jensen NK, Rufford TE, Watson G, Dongke KZ, Chan KI, May EF. Screening zeolites for gas separation applications involving methane, nitrogen, and carbon dioxide. Journal of Chemical & Engineering Data. 2012;**57**(1):106-113. DOI: 10.1021/je200817w

[3] Bae YS, Snurr RQ. Development and evaluation of porous materials for carbon dioxide separation and capture. Angewandte Chemie, International Edition. 2011;**50**:11586-11595. DOI: 10.1002/anie.201101891

[4] Li H, Wang K, Sun Y, Lollar CT, Li J, Zhou HC. Recent advances in gas storage and separation using metal-organic framework. Materials Today. 2018;**21**:108-121. DOI: 10.1016/j.mattod.2017.07.006

[5] Granite EJ, O'Brien T. Review of novel methods for carbon dioxide separation from flue and fuel gases. Fuel Processing Technology. 2005;**86**:1423-1434. DOI: 10.1016/j.fuproc.2005.01.001

[6] Ockwig NW, Nenoff TM. Membranes for hydrogen separation. Chemical Reviews. 2007;**107**(10):4078-4110. DOI: 10.1021/cr0501792

[7] Song L, Sun Z, Duan L, Gui J, McDougall GS. Adsorption and diffusion properties of hydrocarbons in zeolites. Microporous and Mesoporous Materials. 2007;**104**:115-128. DOI: 10.1016/j.micromeso.2007.01.015

[8] Adil K, Belmabkhout Y, Pillai RS, Cadiau A, Bhatt PM, Assen AH, et al. Gas/vapour separation using ultra-microporous metal–organic frameworks: Insights into the structure/separation relationship. Chemical Society Reviews. 2017;**46**:3402-3430

[9] Xu R, Pang W, Yu J, Huo Q, Chen J. Chemistry of Zeolites and Related Porous Materials: Synthesis and Structure. Asia, Singapore: John Wiley & Sons; 2007

[10] Auerbach SM, Carrado KA, Dutta PK. Handbook of Zeolite Science and Technology. New York: Marcel Dekker, Inc.; 2003

[11] Beck DW. Zeolite Molecular Sieves. New York: John Wiley & Sons; 1974

[12] Warrendale PA. Nanoporous and Nanostructured Materials for Catalysis, Sensor, and Gas Separation Applications. San Francisco: Materials Research Society; 2005

[13] Loureiro JM, Kartel MT. Combined and Hybrid Adsorbents: Fundamentals and Applications. Netherlands: Springer; 2006

[14] Schuth F, Sing KSW, Weitkamp J. Handbook of Porous Solids. New York: Wiley-VCH; 2002

[15] Srivastava R, Choi M, Ryoo R. Mesoporous materials with zeolite framework: Remarkable effect of the hierarchical structure for retardation of catalyst deactivation. Chemical Communications. 2006;**43**:4489-4491

[16] Wright PA. Microporous Framework Solids. Cambridge: RSC Publishing; 2008

[17] Corma A. From microporous to mesoporous molecular sieve materials and their use in catalysis. Chemical Reviews. 1997;**97**:23732419. DOI: 10.1021/cr960406n

[18] Corma A. State of the art and future challenges of zeolites as catalysts.

Journal of Catalysis. 2003;**216**:298-312. DOI: 10.1016/S0021-9517(02)00132-X

[19] Choi M, Cho HS, Srivastava R, Venkatesan C, Choi D-H, Ryoo R. Amphiphilic organosilane-directed synthesis of crystalline zeolite with tunable mesoporosity. Nature Materials. 2006;**5**:718-723. DOI: 10.1038/nmat1705

[20] Choi M, Na K, Kim J, Sakamoto Y, Terasaki O, Ryoo R. Stable single-unit-cell nanosheets of zeolite MFI as active and long lived catalysts. Nature. 2009;**461**:246-249. DOI: 10.1038/nature08288

[21] Kim J, Choi M, Ryoo R. Effect of mesoporosity against the deactivation of MFI zeolite catalyst during the methanol-to-hydrocarbon conversion process. Journal of Catalysis. 2010;**269**:219-228. DOI: 10.1016/j.jcat.2009.11.009

[22] Kim J, Park W, Ryoo R. Surfactant-directed zeolite nanosheets: A high-performance catalyst for gas-phase Beckmann rearrangement. ACS Catalysis. 2011;**1**:337. DOI: 10.1021/cs100160g

[23] Wu T, Wang B, Lu Z, Zhou R, Chen X. Alumina-supported AlPO-18 membranes for CO_2/CH_4 separation. Journal of Membrane Science. 2014;**471**:338-346. DOI: 10.1016/j.memsci.2014.08.035

[24] Lee I, Jeong H-K. Synthesis and gas permeation properties of highly b-oriented MFI silicalite-1 thin membranes with controlled microstructure. Microporous and Mesoporous Materials. 2011;**141**:175-183. DOI: 10.1016/j.micromeso.2010.11.012

[25] Michalkiewicz B, Koren ZC. Zeolite membranes for hydrogen production from natural gas: State of the art. Journal of Porous Materials. 2015;**22**:635-646. DOI: 10.1007/s10934-015-9936-6

[26] Das N, Kundu D, Chatterjee M. The effect of intermediate layer on synthesis and gas permeation properties of NaA zeolite membrane. Journal of Coating Technology and Research. 2010;**7**:383-390. DOI: 10.1007/s11998-009-9195-z

[27] Himeno S, Tomita T, Suzuki K, Nakayama K, Yajima K, Yoshida S. Synthesis and permeation properties of a ddr-type zeolite membrane for separation of CO_2/CH_4 gaseous mixtures. Industrial and Engineering Chemistry Research. 2007;**46**:6989-6997. DOI: 10.1021/ie061682n

[28] Li S, Zong Z, James Zhou S, Huang Y, Song Z, Feng X, et al. SAPO-34 membranes for N_2/CH_4 separation: Preparation, characterization, separation performance and economic evaluation. Journal of Membrane Science. 2015;**487**:141-151. DOI: 10.1016/j.memsci.2015.03.078

[29] Sato K, Sugimoto K, Sekine Y, Takada M, Matsukata M, Nakane T. Application of FAU-type zeolite membranes to vapor/gas separation under high pressure and high temperature up to 5 MPa and 180°C. Microporous and Mesoporous Materials. 2007;**101**:312-318. DOI: 10.1016/j.micromeso.2006.12.021

[30] Sebastián V, Kumakiri I, Bredesen R, Menéndez M. Zeolite membrane for CO_2 removal: Operating at high pressure. Journal of Membrane Science. 2007;**292**:92-97. DOI: 10.1016/j.memsci.2007.01.017

[31] Sjöberg E, Barnes S, Korelskiy D, Hedlund J. MFI membranes for separation of carbon dioxide from synthesis gas at high pressures. Journal of Membrane Science. 2015;**486**:132-137. DOI: 10.1016/j.memsci.2015.03.041

[32] Hong M, Li S, Falconer JL, Noble RD. Hydrogen purification using a SAPO-34 membrane. Journal of Membrane Science. 2008;**307**:277-283. DOI: 10.1016/j.memsci.2007.09.031

[33] Tian Y, Fan L, Wang Z, Qiu S, Zhu G. Synthesis of a SAPO-34 membrane on macroporous supports for high permeance separation of a CO_2/CH_4 mixture. Journal of Materials Chemistry. 2009;**19**:7698-7703. DOI: 10.1039/B907237C

[34] Yu M, Li S, Falconer JL, Noble RD. Reversible H_2 storage using a SAPO-34 zeolite layer. Microporous and Mesoporous Materials. 2008;**110**:579-582. DOI: 10.1016/j.micromeso.2007.06.017

[35] Li Y, Yu J. New stories of zeolite structures: Their descriptions, determinations, predictions, and evaluations. Chemical Reviews. 2014;**114**:7268-7316. DOI: 10.1021/cr500010r

[36] Baerlocher C, McCusker LB. Database of zeolite structures. Available from: http://www.iza-structure.org/databases/

[37] Li y LL, Yu J. Applications of zeolites in sustainable chemistry. Chem. 2017;**3**:928-949. DOI: 10.1016/j.chempr.2017.10.009

[38] Mintova S, Gilson JP, Valtchev V. Advances in nanosized zeolites. Nanoscale. 2013;**5**:6693-6703

[39] Bowen TC, Noble RD, Falconer JL. Fundamentals and applications of pervaporation through zeolite membranes. Journal of Membrane Science. 2004;**245**:1-33

[40] Das JK, Das N. Highly oriented improved SAPO 34 membrane on low cost support for hydrogen gas separation. Journal of Materials Chemistry A. 2013;**1**:4966-4973. DOI: 10.1039/c3ta01095c

[41] Himeno S, Tomita T, Suzuki K, Yoshida S. Characterization and selectivity for methane and carbon dioxide adsorption on the all-silica DD3R zeolite. Microporous and Mesoporous Materials. 2007;**98**:62-69. DOI: 10.1016/j.micromeso.2006.05.018

[42] Sen M, Bose A, Pal P, Das JK, Das N. Rapid synthesis of DDR zeolite at room temperature. Journal of the American Ceramic Society. 2014;**97**:52-55. DOI: 10.1111/jace.12687

[43] Zheng Z, Hall AS, Guliants VV. Synthesis, characterization and modification of DDR membranes grown on a-alumina supports. Journal of Materials Science. 2008;**43**:2499-2502. DOI: 10.1007/s10853-008-2560-y

[44] de Man AJM, Van Santen RA. The relation between zeolite framework structure and vibrational spectra. Zeolites. 1992;**12**:269-279

[45] Suslick KS, Doktycz SJ. In: Mason TJ, editor. The Effects of Ultrasound on Solids, in Advances in Sonochemistry. New York, NY: JAI Press; 1990. pp. 197-230

[46] Das JK, Das N, Bandyopadhyay S. Highly selective SAPO 34 membrane on surface modified clay-alumina tubular support for H_2/CO_2 separation. International Journal of Hydrogen Energy. 2012;**37**:10354-10364. DOI: 10.1016/j.ijhydene.2012.03.102

[47] Das JK, Das N. Mercaptoundecanoic acid capped palladium nanoparticles in a SAPO 34 membrane: A solution for enhancement of H_2/CO_2 separation efficiency. ACS Applied Materials & Interfaces. 2014;**6**:20717-20728. DOI: 10.1021/am5045345

[48] Lee JS, Kim JH, Lee YJ, Jeong NC, Yoon KB. Manual assembly of microcrystal monolayers on substrates. Angewandte Chemie, International Edition. 2007;**46**:3087-3090

[49] Yoon KB. Organization of zeolite microcrystals for production of functional materials. Accounts of

Chemical Research. 2007;**40**:29-40. DOI: 10.1021/ar000119c

[50] Bose A, Sen M, Das JK, Das N. Sonication mediated hydrothermal process – An efficient method for the rapid synthesis of DDR zeolite membranes. RSC Advances. 2014;**4**:19043-19052. DOI: 10.1039/c3ra47558a

[51] Das JK, Das N, Roy SN, Bandyopadhyay S. The growth of SAPO 34 membrane layer on support surface for gas permeation application. Ceramics International. 2012;**38**:333-340. DOI: 10.1016/j.ceramint.2011.07.011

[52] Choi J, Ghosh S, Lai Z, Tsapatsis M. Uniformly α-oriented MFI zeolite films by secondary growth. Angewandte Chemie, International Edition. 2006;**45**:1154-1158. DOI: 10.1002/anie.200503011

[53] Boudreau LC, Tsapatsis M. A highly oriented thin film of zeolite A. Chemistry of Materials. 1997;**9**:1705-1709. DOI: 10.1021/cm970151

[54] Caro J, Noack M. Zeolite membranes–recent developments and progress. Microporous and Mesoporous Materials. 2008;**115**:215-233. DOI: 10.1016/j.micromeso.2008.03.008

[55] Zhou M, Liu X, Zhang B, Zhu H. Assembly of oriented zeolite monolayers and thin films on polymeric surfaces via hydrogen bonding. Langmuir. 2008;**24**:11942-11946. DOI: 10.1021/la801879x

[56] Roy P, Das N. Ultrasonic assisted synthesis of Bikitaite zeolite: A potential material forhydrogen storage application. Ultrasonics Sonochemistry. 2017;**36**:466-473. DOI: 10.1016/j.ultsonch.2016.12.032

[57] Munson RA, Clifton RA. Natural Gas Storage with Zeolites. Washington: U.S. Dept. of the Interior; 1971

[58] Berlier K, Olivier MG, Jadot R. Adsorption of methane, ethane, and ethylene on zeolit. Journal of Chemical & Engineering Data. 1995;**40**:1206-1208. DOI: 10.1021/je00022a011

[59] Broom DP. Hydrogen Storage Materials: The Characterization of their Storage Properties. London: Springer-Verlag London; 2011. pp. 19-59

[60] Prasanth KP, Pillai RS, Bajaj HC, Jasra RV, Chung HD, Kim TH, et al. Adsorption of hydrogen in nickel and rhodium exchanged zeolite X. International Journal of Hydrogen Energy. 2008;**33**:735-745. DOI: 10.1016/j.ijhydene.2007.10.047

[61] Zhang SY, Talu O, Hayhurst DT. High-pressure adsorption of methane in zeolites NaX, MgX, CaX, SrX and BaX. The Journal of Physical Chemistry. 1991;**95**:1722-1726. DOI: 10.1021/j100157a044

[62] Cavenati S, Grande CA, Rodrigues AE. Adsorption equilibrium of methane, carbon dioxide, and nitrogen on zeolite 13X at high pressures. Journal of Chemical & Engineering Data. 2004;**49**:1095-1101. DOI: 10.1021/je0498917

[63] Nijkamp MG, Raaymakers JEMJ, van Dillen AJ, de Jong KP. Hydrogen storage using physisorption-materials demands. Applied Physics A: Materials Science & Processing. 2001;**72**:619. DOI: 10.1007/s003390100847

[64] Hassani SS, Salehirad F, Aghabozorg HR, Sobat Z. Synthesis and morphology of nanosized zeolite L. Crystal Research and Technology. 2010;**45**:183-187. DOI: 10.1002/crat.200900450

[65] Regli L, Zecchina A, Vitillo JG, Cocina D, Spoto G, Lamberti C, et al. Hydrogen storage in chabazite zeolite frameworks. Physical Chemistry

Chemical Physics. 2005;7:3197-3203. DOI: 10.1039/B509124A

[66] Kocman V, Gait RI, Rucklidge J. The crystal structure of bikitaite, LiAlSi2O6H2O. American Mineralogist. 1974;**59**:71-78

[67] Palomino GT, Bonelli B, Areá CO, Parra JB, Carayol MRL, Armandi M, et al. Thermodynamics of hydrogen adsorption on calcium exchanged faujasite-type zeolites. International Journal of Hydrogen Energy. 2009;**3**:4371-4378

[68] Li J, Wu E. In: Andreyev MK, Zubkov OL, editors. Storage of Hydrogen in Zeolites in Zeolites: Synthesis, Chemistry and Applications. New York: Nova Science Publishers, Inc; 2012. pp. 149-170

[69] Roy P, Das N. Synthesis of NaX zeolite-graphite amine fiber composite membrane: Role of graphite amine in membrane formation for H_2/CO_2 separation. Applied Surface Science. 2019;**480**:931-944. DOI: 10.1016/j.apsusc.2019.03.038

Chapter 7

Zeolites Applications in Veterinary Medicine

Marc Simona and Tulcan Camelia

Abstract

Zeolites have a wide range of use, from construction industries, aquaculture industries, agriculture, space research to human and veterinary medicine. This broad application of natural and synthetic zeolites is given by their main properties: adsorption, molecular sieving and cation exchange capacity. In this chapter the main use of zeolites in veterinary medicine is reviewed. The beneficial effects of zeolites in animal nutrition, on mycotoxins, as an adjuvant in anticancer treatment and in increasing passive immunity of newborn ruminants are reported. Furthermore, multiple advantageous immune effects of zeolites such as their antioxidant capacity or their non-specific superantigen-like immunoglobulin action are also reviewed. Finally, their main positive effect on passive immunity in newborn calves is discussed. Literature data reviewed confirms their beneficial role in newborn calves during colostral period.

Keywords: zeolites, feed additive, immunostimulation, ruminants

1. Introduction

The first group of zeolite minerals was discovered in 1765 by a Swedish mineralogist A.F. Cronstedt who described several species found in well-defined crystals. He noticed that some heated minerals began to lose their constituent water with a boiling-like appearance, hence the name of zeolite (from the Greek words "zeos" and "lithos" which translate as "boiling stone") [1].

Zeolites are natural, hydrated, crystalline aluminosilicates made up of three-dimensional networks extended by AlO_4 and SiO_4 linked together by oxygen atoms, which make up a rigid, open, honeycomb-like skeleton, generally including cations which neutralize the excess negative charge of aluminum tetrahedra and water molecules. Each AlO_4 tetrahedral network supports a net negative charge that is balanced by a cation, usually from the I-A or II-A group (Ca, Mg, Na, K, Fe). These ions are not part of the zeolite network and can be changed by other cations such as heavy metals (Hg, Pb, Cd) or ammonium ions [2–4].

There are 67 natural zeolite minerals accepted by the Natural Zeolites Commission of the International Zeolite Association (IZA) and all have a unique three-letter code [5].

Clinoptilolite of sedimentary origin, generally the most used natural zeolite, is authorized by the Commission Implementing Regulation (EU) no. 651/2013 as a feed additive for all animal species [6]. In the United States, clinoptilolite belongs to the sodium aluminosilicate category and has the status of Generally Recognized as Safe (GRAS) (Code of Federal Regulations CFR, Title 21, Section 182.2727) [7].

Due to their main properties: adsorption, molecular sieving and cation exchange, zeolites have a wide use in many areas. For example, in agriculture, natural zeolites are used to obtain fertilizers capable of better nitrogen retention and in a slow and controlled release of fertilizers, nitrogen use efficiency (NUE) increase [8, 9]. In aquaculture industry natural, synthetic or modified zeolites are used as adsorbents for ammonia removal from fish farming ponds and transportation tanks, as a cation-exchanger for removal of different toxic heavy metals from fresh water and sea water cultures and as a feed supplement to enhance fish growth [3]. Also, zeolites can increase the nutrient (by addition of micronutrients) and water use efficiency of drylands (by their water holding capacity) [9]. Natural zeolites are used in wastewater, surface waters, ground and underground water, drinking water treatment [10, 11], in decontamination of radioactive waste water [12] and in agro-industrial wastewater treatment due to their exceptional cationic exchange and adsorption properties [13]. In construction field through their excellent properties, mainly porosity, specific weight and adsorption, they can be used as a building stone [14], in zeoponic substrates—artificial soils developed by the National Aeronautics and Space Administration—for plant growth in space [15], and as potential slow-release carriers for herbicides, insecticides and other organic compounds, protecting in this way the environment from chemicals [16].

In animal production, alternative products as zeolites are a solution to ensure health, productive performance (yield and quality of carcass, milk yield), to reduce the effects of mycotoxins on animal health status, to remove selectively pathogens from the animal gut without reducing microbial richness and finally to increase farm profitability. All of these effects has been extensively studied in the last decade [17–28] and are schematic represented in **Figure 1**.

Clinoptilolite is also used as a biomedical feed ingredient due to its beneficial properties as a growth-promoter and immunostimulant and can constitute an alternative to antibiotic growth promoters [29], since European Union legislation has banned the use of antibiotics for growth promotion in 2006, because the overuse of antibiotics in animals can contribute to emergence of antibiotic resistance [30].

Zeolites can have a protective effect in intoxication and in reducing parasite infestations. These effects are evidenced by researchers who observed that clinoptilolite (2 g/kg) could have some protective effect in organophosphorus poisoning in sheep by protecting the rumen flora and by preventing the decrease of cholinesterase activity [31]; in lead intoxication in mice, clinoptilolite given in 10/1 ratio

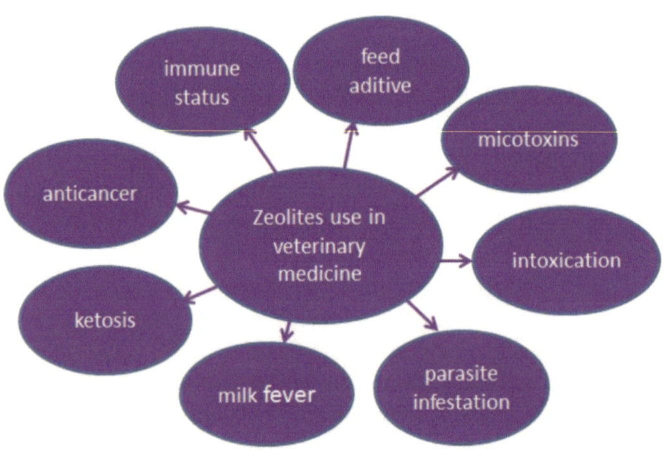

Figure 1.
The main applications of zeolites in veterinary medicine.

(clinoptilolite/Pb) adsorbed 91% of Pb, and when supplementing 3% clinoptilolite feed to pigs that received 150 ppm $CdCl_2$, clinoptilolite prevented Cd-induced anemia by adsorption of Cd [32]; and in reducing the excretion of cysts in goat kids with giardiasis [33]. It also had beneficial effects in infestations with nematode larvae in lambs, producing an increase in feed consumption and in body mass [34].

Some types of zeolites are studied for their adsorption properties in order to improve the life of people suffering from chronic kidney disease, who need to undergo weekly hemodialysis. Dialysis membranes made from zeolite and polymer are studied in order to improve the performance of hemodialysis. Nanofiber membranes made from zeolite and polyacrylonitrile (PAN) adsorb creatinine, with the best results for 940-HOA (beta) zeolites (25,423 μg/g in 625 μmol/L creatinine solution) [35] and P87 zeolite in combination with polyethersulfone, used because of their improved resistance to fouling, thermal stability, chemical resistance and due to their high adsorption level of indoxyl sulfate (550 μg/g membrane) [36]. When zeolite was used along polyethersulfone (PES) and dimethyl formamide (DMF) in proper concentration: 17:0.5:82.5 (PES:zeolit:DMF), creatinine concentration decreased by 91.99%, which suggests the possibility of using these membranes in haemodialysis [37].

2. As dietary supplements in animal nutrition

In animal husbandry, natural and synthetic zeolites have been mostly used to improve productive performance. The proposed mechanisms involved in achieving the increase in productive performance in animals are: ammonia binding, reducing toxic effects of ammonia produced by intestinal microbial activity; low passage rate of digesta through the intestines and more efficient use of nutrients; enhanced pancreatic enzymes activity-favorable effect on feed components hydrolysis over a wider range of pH, improved energy and protein retention; elimination of mycotoxin growth inhibitory effects [32].

Due to the beneficial effects of the gradual release of ammonia ions on microbial synthesis in the rumen, zeolites are used especially in high non-protein nitrogen feed ratio. *In vitro* and *in vivo* experimental studies have shown that 15% of ruminal NH_4^+ can be adsorbed by zeolites, thus reducing the toxic effects of urea (increased rumen pH and ammonia concentration in rumen and blood). Thus clinoptilolite (6%) in the feed of dairy cows receiving urea significantly reduced the concentration of NH_4^+ in the rumen [32]. Also, a decrease in ruminal pH in diets with 1% clinoptiloliten is reported [38].

Milk fever and ketosis are the most common metabolic diseases that occur in cows with high milk production. Cows that received zeolite (1 kg zeolite/day for 4 weeks before calving) did not experience subclinical hypocalcemia [39]. Also, the administration of zeolite A (sodium aluminosilicate) to pregnant cows during the dry period (1.4 kg zeolite A/day in the last 2 weeks of gestation) reduced the incidence of milk fever. The mode of action of synthetic zeolite A is to reduce the bioavailability of fodder calcium at the gastrointestinal level (calcium binding capacity of zeolite is 110 mg/g Ca at pH 11), stimulating Ca-homeostatic mechanisms before calving. At calving, the plasma level of calcium was significantly higher in the experimental group ($p < 0.0001$); with a slight drop of inorganic magnesium and phosphorus, that set up a week postpartum [40].

When clinoptilolite was administered (2.5%) in the last month of gestation, the incidence of milk fever was 5.9%, compared to 38.9% in the control group. Also, clinoptilolite (2.5%) administered during the dry period reduced the incidence of ketosis (5.9%) by improving the energy metabolism through increased production of propionate in the rumen and by better recovery of feed [32]. Katsoulos et al. revealed that long-term clinoptilolite administration (from 4 weeks before calving to the next dry period) at different doses (1.25 and 2.5%) did not have adverse effects on the liver and serum glucose concentrations, ketone bodies, total protein and urea did not change, with a higher milk production and a lower ketosis incidence [41]. Moreover, other important minerals such as: Cu, Zn, Fe were not influenced by the long-term administration of clinoptilolite (1.25 and 2.5%), which highlights the safety of this natural compound [19].

Also, in combination with yeast, clinoptilolite (Rumencure: yeast 60% and clinoptilolite 40%) given to cows for a long period (30 days) had no apparent adverse effects on their liver function and on some biochemical parameters (glucose, ketone bodies, blood urea nitrogen and total proteins) [42].

3. Positive effects against mycotoxins

Due to the increased incidence of contamination with mycotoxins, it has been attempted to use inert feed adsorbents to bind mycotoxins, thus reducing their

Dietary rate	Species (N)	Effect	Ref.
20 g/kg clinoptilolite in feed contaminated with 1 mg/kg aflatoxins for 42 days	Broiler chickens N = 480	Decreased the severity of lesions and effectively diminished the detrimental effects of aflatoxins	[45]
1% synthetic zeolites NaA in feed contaminated with 2.5 mg/kg aflatoxin B1 from 21 to 42 days of age	Male broiler chicks N = 80	Zeolite NaA can counteract some of the toxic effects of aflatoxin A in growing broiler chicks	[46]
3 and 5% clinoptilolite in feed contaminated with 2 ppm aflatoxin from day 1 to 7 weeks of age	Male chicks N = 900	The level of 5% clinoptilolite was better in reduction the effects of alfatoxin than 3% clinoptilolite ratio	[47]
0.2% Minazel Plus/0.2% Mycosorb/0.2% Mycofix-plus in feed contaminated with 2 ppm T-2 toxin for 21 days	"Ross" broiler chicks N = 160	Pathohistological examination of liver, bursa of Fabricius and small intestine revealed better protective effects in groups fed with Mycofix-plus than in groups with Minazel Plus and Mycosorb were protective failure was noted	[48]
2% clinoptilolite in feed contaminated with 2.5 ppm aflatoxin B1 for 4 weeks	Laying hens N = 96	The livers of hens showed very low mycotoxin concentrations	[49]
0.2% Min-a-Zel Plus in feed contaminated with 3 mg/kg zearalenone for 14 days	Piglets N = 20	Agonistic effect due to oestrogen reduction	[22]
0.2% organozeolite 0.5% organozeolite in feed contaminated with 8.3 mg ZEN/kg for 53 days	Lambs N = 64	The organozeolite reduced the content of zearalenone in liver, kidneys and muscles Addition of 0.5% Min-a-Zel Plus eliminated zearalenone from all organs, totally	[17]
200 g clinoptilolite/animal/day for 7 day	Dairy cattle N = 15 farms	Significantly reduced aflatoxin M1 in milk at an average rate of 56.2%	[50]

Table 1.
The summarized effects of zeolites on mycotoxins as reported in literature data.

intestinal absorption and toxic effects on animals and animal products. Annually, it is estimated that about 25% of the world's harvested crops are contaminated by mycotoxins, leading to huge agricultural and industrial losses [43]. The first adsorbents successfully used in poultry, swine, sheep and bovine breeding were phyllosilicates, namely bentonite [32].

The adsorption process is strongly related to the pore size, the adsorbent contact surface, polarity, solubility, and the size of the mycotoxin molecules that are adsorbed (e.g., aflatoxins B1 and B2 have 5.18 Å and aflatoxins G1 and G2 are 6.50 Å). Clinoptilolite has the highest *in vitro* adsorption, over 80% for aflatoxins B1 and G2 [44], with effects demonstrated especially in the poultry industry [45–49], but also in piglets [22], lambs [17] and dairy cattle [50] as are presented in **Table 1**.

Studies performed by Serbian researchers have demonstrated *in vivo* and *in vitro* that clinoptilolite preparations adsorb ohratoxin A, zearalenone, aflatoxin B1, B2, G2, T-2 toxin, ergosine, ergocristine, ergocryptine and ergometrine in feed [51, 52]. The proportion of adsorption by clinoptilolite particles of the toxins enumerated *in vitro* varies depending on the concentration of these toxins and can range up to 99%. The main mycotoxins adsorbed by Min-a-Zel Plus, modified clinoptilolite, are: aflatoxin B1–99%, zearalenone—94%, ochratoxin A—96% and ergot alkaloids—97% [51, 52]. In another study, T2 toxin—a secondary metabolite of *Fusarium fungi*—was adsorbed on average in 30% by Min-a-zel Plus, Mycosorb (esterified glucomanane) and mixed binder Mycofix (inorganic binder, bacteria, enzymes and phytogenic material extracted from plants) in *in vitro* conditions at pH 3 [53].

4. Adjuvant in anticancer treatment

The first studies of anticancer effect of zeolites were performed at the beginning of 2001 when it was observed that the treatment with clinoptilolite of different tumors in mice and dogs have improved their life span and tumors have decreased in size [54]. *In vitro* studies using cancer cell cultures revealed the clinoptilolite inhibitory effect on protein kinase B (B-Akt), which reduced the growth of cancer cells and increased their apoptosis. Inhibition occurred only in the presence of serum. This finding suggests that adsorption of serum components may be a possible mode of action. Adsorption of molecules involved in transduction signals, such as inositol, phosphatides and Ca, can contribute to its therapeutic efficacy. It also induces expression of tumor suppressor proteins, p21 WAF1/CIP and p27 KIP1, blocking the growth of cancer cell lines. It is assumed that clinoptilolite reduces the exchange rate of intestinal epithelial cells, prolonging their activity, and that silicates and aluminosilicates can interact directly with some cells by modifying their intracellular pathways, and this leads to the alterations in the regulation of gene expression. Changing the order of interaction of other proteins with membrane proteins may be involved, since membrane transport is required to activate protein kinase B [54]. Studies in mice injected i.v. with melanoma cells but receiving a micronized zeolite through gastric intubation for 28 days, revealed an increase in allogeneic graft versus host (GVH) in lymphocytes in the lymph nodes and a reduction in pulmonary metastases. The researchers' hypothesis is that the local inflammation caused by zeolite application, attracts peritoneal macrophages, and these cells in turn produce TNFα that stimulates spleen T-cells, which amplify the local inflammatory response [55]. Also, a reduction in the metabolic rate of cancer cells and a reduced production of 4-hydroxyinonenal following an anti-cancer treatment (Doxorubicin) along with tribomechanically micronized clinoptilolite, having in this way a potentiator effect on anticancer drugs, were reported [56].

In another *in vitro* study done on mouse fibrosarcoma cells and other types of cells incubated for 24 h together with clinoptilolite researchers observed that the number of viable cells, DNA synthesis and activity of EGF-R, PKB/Akt and NFKB was reduced while apoptosis was enhanced maybe because clinoptilolite affects cellular microenvironment through mechanisms that are dependent on its characteristics [57].

5. Effects on health status and growth performance

Clinoptilolite is also used as feed ingredient due to its beneficial properties as immunostimulant. One explanation of beneficial immune effects of silica, silicates and aluminosilicates could be their action as non-specific superantigen-like immunoglobulins (SAg). SAg are viral and bacterial toxins that are capable of activating a large population of T-cells. Activation occurs as a result of the simultaneous interaction between SAg, the T cell receptor (TcR) variable region β and the major histocompatibility complex (MHC) class II molecules on the surface of antigen presenting cells (APC). Consequently, SAg stimulates 10–30% of T-cells, as opposed to 0.01–0.0001% as it stimulates common antigens. Proinflammatory macrophages belonging to APC cells, CMH class II are activated by the particles of silicates [58].

An indirect action of clinoptilolite on the immune system is also achieved by its antioxidant capacity. Sverko et al. showed that administration of tribomechanically-activated clinoptilolite (12.5%) alone or together with *Urtica dioica* extract in mice *per os* for 3 weeks significantly reduced lipid peroxidation processes in the liver and significantly increased the content of superoxide-dismutase, an antioxidant enzyme. The antioxidant role of clinoptilolite is probably given by positive electrons that neutralize free radicals [59].

In weaned piglets that received 0.5% clinoptilolite for 5 weeks, it was observed that clinoptilolite was effective as an immunomodulatory agent by promoting the recruitment of circulating and intestinal immune cell subsets, even though it did not improve growth in weaned pigs, and generally failed to improve their feed conversion efficiency [60]; in heifers vaccinated against *Escherichia coli* (day 210 and 240 of gestation) the potentiating effect of clinoptilolite on the immune response was highlighted, and it indirectly improved the protection of calves [61].

In newborns, adding clinoptilolite to colostrum improves intestinal absorption of colostrum globulin, creating a good protection against neonatal diseases [18, 62–64].

Colostral period represents an important moment for the newborn ruminants, because in these species acquiring passive immunity is achieved exclusively through ingestion and absorption of adequate amounts of colostral immunoglobulins (Ig) [65]. Obtaining a good protection against neonatal diseases depends on how well this period is managed.

Scientific evidence highlights positive effects of zeolite supplementation on passive immunity [18, 21, 62] and on biochemical parameters [20, 23, 66] in newborns calves.

When we analyzed the effects of clinoptilolite supplemented in colostrum on blood serum protein electrophoretic pattern of newborn calves that received colostrum supplemented with 0.5 and 2% clinoptilolite, at 30 h after birth, the concentrations of γ-globulins, β-globulin and total protein in the group of calves that received 0.5% clinoptilolite (E1) were higher than in the control group by 42.11% ($p < 0.05$), 28.48% ($p > 0.05$) and 18.52% ($p > 0.05$), respectively, and were higher, but not significantly, in the group that received 0.5% clinoptilolite (E2), and a significantly lower albumin/globulin ratio in groups E1 and E2 (29.35%, $p < 0.05$

and 35.87%, $p < 0.05$, respectively) was reported, compared with the control group at 30 h postpartum, which indicates an obvious increase in globulin fractions in experimental groups. Clinoptilolite was effective in improving passive transfer in newborn calves, better if added in colostrum in a dose of 0.5% than in a dose of 2% (**Figure 2**) [67].

A possible way of clinoptilolite action is explained, based on observation in Wistar rats that received zeolite for 34 days (6% of their weight) [68]. Some modifications of intestinal cells were observed, such as: the microvilli length was higher (1.2 vs. 2.0), the number of microvilli per 10 μm was higher (54.4 vs. 64.8) their diameter was smaller (0.17 vs. 0.13) compared to the control group, and also that the cellular organisms of the enterocytes, the density of mitochondrial membranes and the number of attached ribosomes were higher, which indicates a rise in the adaptation processes of the cells [68].

In an experiment carried out in 20 newborn calves in order to observe the clinical effects of clinoptilolite added in colostrum (20 ml clinoptilolite/L colostrum) during the first three meals, we concluded that administration of zeolites appears to reduce the incidence of diarrhea because only two calves from experimental group had health problems, one had bronchopneumonia and the other digestive transit difficulties with symptoms appearing after 28 days (not in neonatal period) in comparison with the control group where all calves had diarrhea in the first 11 days of life. The other parameter, growth performance measured on day 0, 45 and 90 revealed that during the first 45 days body weight of calves treated with clinoptilolite was significantly higher compared with the group of calves receiving only colostrum (C) ($p < 0.0058$) (E1/C = +16.96%). This statistical difference at 45 days may be explained by the high number of calves from the control group suffering from health problems and this affected the daily gain. At 90 days, the difference between groups was not significant ($p < 0.1035$) (E2/C = +7.19%) [69, 70]. Similar results were obtained by Step et al. who found that body weight and average daily gain did not differ between treatment groups (clinoptilolite dosage was 0.5 and 2%) [23]. More

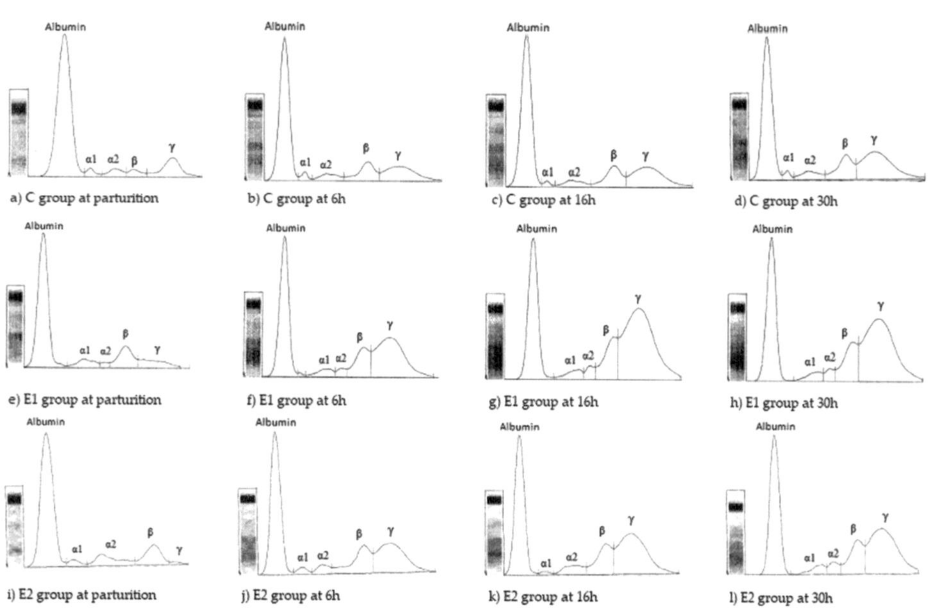

Figure 2.
Representative serum protein electrophoretograms observed in calves that received colostrum supplemented with 0.5% (e–h) and 2% clinoptilolite (i–l) compare with control group (a–d) at different time interval (0, 6, 16 and 30 h postpartum) [67].

recently, Ural et al. observed increased total weight and mean daily gain in calves that receive clinoptilolite (1 or 2 g/kg) in colostrum at calving, 12 and 24 h [71].

In another study, the addition of 0.5 g and 1 g/kg body weight per day in colostrum and milk for 45 days reduced fecal score and its severity, probably by retarding effect of clinoptilolite on intestinal passage rate [21]. The activity of clinoptilolite on reducing signs of diarrhea could be caused by: alteration of metabolic acidosis through effects on osmotic pressure in the intestinal lumen; or through retention of the enterotoxigenic *E. coli* thus limiting its attachment to the intestinal cell-membrane receptors); and also due to water adsorption property of zeolites, the feces appear drier and more compact [32].

In human medicine there are studies that support the beneficial properties of purified natural clinoptilolite as an anti-diarrheic treatment [72]. More recent studies performed on aerobically trained subjects, who received for 12 weeks zeolite-clinoptilolite supplementation, highlighted the positive effects of zeolites on intestinal wall integrity. The results were based on decreased concentrations of zonulin, an intercellular tight junction modulator, improving in this way intestinal barrier integrity [73].

Also, clinoptilolite improve antioxidant capacity in broilers [74]; it is used as a feed additive in fish diets [75] and in turkey diets [76].

Analyzing the effects of clinoptilolite on mineral parameters in newborn Romanian Black and White calves that received 5 g/l (group E1) and 20 g/l (group E2) at parturition, and 12 and 24 h postpartum, we observed that clinoptilolite supplementation increased serum Ca (with 37.34% in group E1, with 21.42% in group E2 in comparison with the control group and with 13.11% in E1/E2 at 30 h postpartum), P (with 37.34% in group E1, with 21.42% in group E2 in comparison with the control group and with 13.11% in E1/E2 at 30 h postpartum), Mg concentration (increased in groups E1 in comparison with the control group ($p < 0.003$) and E2 ($p < 0.009$) at 30 h postpartum in neonatal calves) with the most spectacular increase in iron concentration (with +144.70% in group E1 ($p < 0.0005$) and with +126.16% in group E2 ($p < 0.002$) at 30 h postpartum) [64].

When analyzing the same parameters in other breed (Holstein) and other colostrum quantity (3 L), we observed that the most significant effect ($p < 0.0006$) was on serum iron concentrations in experimental (27.64 ± 3.78 μmol/l) vs. control group (8.93 ± 1.26 μmol/l) and it did not have negative effects on other biochemical parameters (Ca, P, Mg, GGT, ALAT, ASAT, ALP) after 48 h postpartum, one more time proving that morpho-functional processes that take place in the newborns, necessary for adapting to the new environment, were not affected by clinoptilolite [63]. A possible explanation of increased iron level could be that in duodenum and in the anterior part of jejunum, where iron absorption takes place, clinoptilolite influences iron absorption due to the ion exchange properties, altering in this way the pH or reducing intestinal transit of digesta, which could lead to a better utilization of nutrients [32, 77]. It has been shown previously that low intestinal motility and acidic pH promote iron absorption and that in the bovine neonate, the pH of the whole intestinal content ranges from 5.5 to 6.5; also, the motility of the gastrointestinal tract becomes well organized only after 2–3 days of postnatal life [78, 79]. This feature could be important in preventing iron deficiency anemia (Fe < 14.32 μmol/l) especially in veal calves fed exclusively with milk.

Short term supplementation of clinoptilolite did not affect hepatic and renal function of newborn calves and that morpho-functional changes of the newborn organism in adapting to extrauterine environment were normal, without any influence of clinoptilolite, as observed after analyzing α-amylase, total bilirubin, creatinine, uric acid, urea, glucose, cholesterol and triglycerides. Biochemical values were measured in the first 48 h in the newborn calves. Values recorded were physiological for the neonatal period and had no significant difference between groups, highlighting once more the safety of clinoptilolite in newborn nutrition [80].

The activity of the most important enzymes is changing very fast after the first feedings. We observed that adding clinoptilolite in the first three meals of colostrum influenced the enzymes as follows: GGT activity significantly increased in group E2 (20 g/L clinoptilolite) at 6 h after birth (E2/C: +64.83%, $p < 0.05$) and in group E1 (5 g/L clinoptilolite) at 16 h after birth (E1/C: +118.55%, $p < 0.05$) in comparison with the control group (C—received only mother colostrum); ALP increased after birth in all calves and adding clinoptilolite in colostrum influenced activity of ALP only in group E1 at 30 h postpartum; transaminases were low at birth in all calves but after feeding they increased, this coincided with the period when a morpho-functional condition of the liver is changing in a newborn calf; adding clinoptilolite (5 g/L colostrum) to colostrum determined increased ASAT (E1/C: +71.58%, $p < 0.01$) and ALAT (E1/C: +278.82%, $p < 0.006$) activity at 30 h postpartum [81]. As literature data suggests, serum GGT is the only enzyme to increase markedly as a result of its absorption from the colostrum; other serum enzymes, such as aspartate aminotransferase (ASAT) and alkaline phosphatase (ALP), are presumably released from the tissues of the calf [82]. A good interpretation of the serum enzyme activity in newborn calves must consider the physiological increase which occurs after feeding colostrum in the first days after parturition, a period very important for the calf but also for the cow [83].

6. Conclusion

Based on our research data and on continuously published literature data worldwide regarding the use of zeolites in veterinary medicine, we confirm that they can be used in animal nutrition as feed additives, mainly to reduce the gastrointestinal absorption of mycotoxins; in newborn calves, they can be used as enhancers of passive immunity during colostral period; also, to increase health status and growth performance of animals and as an adjuvant in anticancer treatment, with a promising perspective in this field.

Acknowledgements

This work was performed through the project “Ensuring excellence in R&D activity within BUASVM Timisoara” code 35PFE, developed with the support of Ministry of Research and Innovation, Romania, the contracting authority for Program 1—Developing national R&D, Subprogram 1.2—Performance Institutional, Institutional Development Project—Financing projects of excellence in R&D.

Author details

Marc Simona*† and Tulcan Camelia†
Faculty of Veterinary Medicine, Banat's University of Agricultural Sciences and Veterinary Medicine "King Michael I of Romania" from Timişoara, Romania

*Address all correspondence to: simo20_med@yahoo.com

† Equal contribution authorship.

References

[1] Zimmermann NER, Haranczyk M. History and utility of zeolite framework-type discovery from data-science perspective. Crystal Growth and Design. 2016;**16**:3043-3048. DOI: 10.1021/acs.cgd.6b00272

[2] Georgiev D, Bogdanov B, Angelova K, Markovska I, Hristov Y. Synthetic zeolites—Structure, classification, current trends in zeolite synthesis review. In: International Science Conference "Economics and Society Development on the Base of Knowledge"; 4-5 June 2009; Stara Zagora, Bulgaria; Technical Studies; volume VII; pp. 1-5

[3] Mumpton FA. La roca magica: Uses of natural zeolites in agriculture and industry. Proceedings of the National Academy of Sciences. 1999;**96**:3463-3470

[4] Roth WJ, Nachtigall P, Morris RE, Cejka J. Two-dimensional zeolites: Current status and perspectives. Chemical Reviews. 2014;**114**:4807-4837. DOI: 10.1021/cr400600f

[5] International Zeolite Association [Internet]. Available from: http://www.iza-online.org/ [Accessed: 06 April 2019]

[6] European Union. Authorisation of Clinoptilolite Use as a Feed Additive [Internet]. Available from: https://www.ecolex.org/details/legislation/commission-implementing-regulation-eu-no-6512013-concerning-the-authorisation-of-clinoptilolite-of-sedimentary-origin-as-a-feed-additive-for-all-animal-species-and-amending-regulation-ec-no-18102005-lex-faoc125758/ [Accessed: 07 April 2019]

[7] FDA. U.S. Food and Drug Administration [Internet]. Available from: https://www.accessdata.fda.gov/scripts/cdrh/cfdocs/cfcfr/CFRSearch.cfm?fr=182.2727 [Accessed: 06 April 2019]

[8] Bernardi AC, Mota EP, SCH S, Cardoso RD, Peronti P, Oliviera A. Ammonia volatilization, dry matter and nitrogen levels of Italian ryegrass fertilized with urea and zeolite. In: 19th World Congress of Soil Science, Soil Solutions for a Changing World; 1-6 August. 2010. pp. 22-25

[9] Jakkula VS, Wani SP. Zeolites: Potential soil amendments for improving nutrient and water use efficiency and agriculture productivity. Scientific Reviews & Chemical Communications. 2018;**8**(1):119

[10] Karmen M, Natasa ZL, Siljeg M, Anamarija F. Natural Zeolites in Water Treatment—How Effective Is their Use. Intech; 2013. DOI: 10.5772/50738

[11] Wang S, Peng Y. Natural zeolites as effective adsorbents in water and wastewater treatment. Chemical Engineering Journal. 2010;**156**(1):11-24. DOI: 10.1016/j.cej.2009.10.029

[12] Yeritsyan H, Sahakyan A, Harutyunyan V, Nikoghosyan S, Hakhverdyan E, Grigoryan N, et al. Radiation-modified natural zeolites for cleaning liquid nuclear waste (irradiation against radioactivity). Scientific Reports. 2013;**3**:2900. DOI: 10.1038/srep02900

[13] Kotoulas A, Agathou D, Triantaphyllidou I, Tatoulis TI, Akratos CS, Tekerlekopoulou AG, et al. Zeolite as a potential medium for ammonium recovery and second cheese whey treatment. Water. 2019;**11**(1):13. DOI: 10.3390/w11010136

[14] Samardzioska T, Jovanovski M. Zeolites-sustainable building material. In: Proceedings of the 1st International Conference on

Construction Materials for Sustainable Future; Zadar, Croatia; 19-21 April. 2017. pp. 146-151

[15] Ming DW, Allen ER. Zeoponic substrates for space applications: Advances in the use of natural zeolites for plant growth. Natural Microporous Materials in Environmental Technology. 1999;**362**. DOI: 10.1007/978-94-011-4499-5_11

[16] Bish DL, Ming DW. Natural zeolites: Occurrence, properties, applications. Mineralogical Society of America, Reviews in Mineralogy. 2001;**45**:639. ISBN 1529-6466

[17] Dakovic A, Adamovic M, Tomasevic-Canovic M. Efficacy of organozeolite to ameliorate the toxic effects of zearalelone in lambs. Acta Veterinaria. 2004;**54**(1):53-62

[18] Fratric N, Stojic V, Jankovic D, Samanc H, Gvozdic D. The effect of a clinoptilolit based mineral adsorber on concentrations of immunoglobulin G in the serum of newborn calves fed different amounts of colostrum. Acta Veterinaria. 2005;**55**(1):11-21

[19] Katsoulos PD, Roubies N, Panousis N, Karatzias H. Effects of long-term feeding dairy cows on a diet supplemented with clinoptilolite on certain serum trace elements. Biological Trace Element Research. 2005;**108**(1-3):137-145. DOI: 10.1385/BTER:108:1-3:137

[20] Mohri M, Seifi HA, Maleki M. Effects of short-term supplementation of clinoptilolite in colostrum and milk on the concentration of some serum minerals in neonatal dairy calves. Biological Trace Element Research. 2008;**123**(1-3):116-123

[21] Sadeghi A, Shawrang P. Effects of natural zeolite clinoptilolite on passive immunity and diarrhea in newborn calves. Livestock Science. 2008;**113**(2-3):307-310. DOI: 10.1016/j.livsci.2007.08.010

[22] Speranda M, Liker B, Speranda T, Seric V, Antunovoc Z, Grabarevic Z, et al. Hematological and biochemical parameters of weaned piglets fed on fodder mixture contaminated by zearalenone with addition of clinoptilolite. Acta Veterinaria. 2006;**56**(2-3):121-136

[23] Step DS, Litherland NB, Burciaga-Roble O, Breshears M, Krehbiel CR, Confer AW, et al. Clinical observations, biochemical data and postmortem and histopathologic findings in young dairy calves fed zeolite clinoptilolite binder combined with milk replacer. American Journal of Veterinary Research. 2008;**69**(12):1587-1594

[24] Stojic V, Gvozdic D, Anna N, Samanc H, Jovanovic I, Magdalena T-C, et al. The serum levels of insulin and IGF-1 in newborn piglets treated with clinoptilolite. Acta Veterinaria. 2003;**53**(4):219-228

[25] Prasai TP, Walsh KB, Bhattarai SP, Milmore DJ, Van TTH, Moore RJ, et al. Biochar, bentonite and zeolite supplemented feeding of layer chickens alters intestinal microbiota and reduces Campylobacter load. PLoS One. 2016;**11**(4). DOI: 10.1371/journal.pone.0154061

[26] Schneider AF, Zimmermann OF, Gewehr CE. Zeolites in poultry and swine production. Animal Production. 2017;**47**(8). DOI: 10.1590/0103-8478cr20160344

[27] Burmańczuk A, Roliński Z, Kowalski C, Burmańczuk N, Markiewicz W. Possibilities of using the natural zeolites in animals production and environment protection. Journal of Elementology. 2014;**20**(4):803-811. DOI: 10.5601/jelem.2014.19.4.759

[28] Khachlouf K, Hamed H, Gdoura R, Gargouri A. Effects of zeolite supplement on dairy cow production and ruminal parameters—A review. Annals of Animal Science. 2018;**18**(4):857-877. DOI: 10.2478/aoas-2018-0025

[29] Valpotic H, Gracner D, Turk R, Duricic D, Vince S, Folnozic I, et al. Zeolite clinoptilolite nanoporous feed additive for animals of veterinary importance: Potentials and limitations. Periodicum Biologorum. 2017;**119**(3):159-172. DOI: 10.18054/pb.v119i3.5434

[30] World Health Organization [Internet]. Available from: https://www.who.int/news-room/detail/07-11-2017-stop-using-antibiotics-in-healthy-animals-to-prevent-the-spread-of-antibiotic-resistance [Accessed: 07 April 2019]

[31] Nistiar F, Mojžiš J, Kovac G, Seidel H, Rácz O. Influence of intoxication with organophosphates on rumen bacteria and rumen protozoa and protective effect of clinoptilolite-rich zeolite on bacterial and protozoan concentration in rumen. Folia Microbiologica. 2000;**45**:567-571

[32] Papaioannou D, Katsoulos PD, Karatzias H. The role of natural and synthetic zeolites as feed additives on the prevention and/or the treatament of certain farm animal diseases: A review. Microporous and Mesoporous Materials. 2005;**84**:161-170. DOI: 10.1016/j.micromeso.2005.05.030

[33] Ural DA, Erdoğan E, Toplu S, Ayan A. Application of oral clinoptilolite intended for prevention of giardiasis in goat kids. Kocatepe Veterinary Journal. 2017;**10**(3):158-163

[34] Deligiannis K, Lainas TH, Arsenos G, Papadopoulos E, Fortomaris P, Kufidis D, et al. The effect of feeding clinoptilolite on food intake and performance of growing lambs infected or not with gastrointestinal nematodes. Livestock Production Science. 2005;**96**(2-3):195-203

[35] Limin L, Champika S, Yeow JTW. Creatinine adsorption capacity of electrospun polyacrylonitrile (PAN)—Zeolite nanofiber membranes for potential artificial kidney applications. Journal of Applied Polymer Science. 2015;**132**:42418. DOI: 10.1002/app.42418

[36] Limin L, Yeow JTW. An adsorption study of indoxyl sulfate by zeolites and polyethersulfone-zeolite composite membranes. Materials & Design. 2017;**120**:328-335. DOI: 10.1016/j.matdes.2017.01.094

[37] Prelina B, Wardana J, Isyatir RA, Syukriyah Z, Wafiroh S, Raharjo Y, et al. Innovation of zeolite modified polyethersulfone hollow fibre membrane for haemodialysis of creatinine. Chemistry & Chemical Technology. 2018;**12**(3):331-336. DOI: 10.23939/chcht12.03.331

[38] Mumpton FA, Fishman PH. The application of natural zeolites in animal science and aquaculture. Journal of Animal Science. 1977;**45**:1188-1203

[39] Thilsing-Hansen T, Jorgensen RJ. Prevention of parturient paresis and subclinical hypocalcemia in dairy cows by zeolite A administration in the dry period. Journal of Dairy Science. 2001;**84**(3):691-693

[40] Thilsing-Hansen T, Jorgensen RJ, Enemark JMD, Larsen T. The effect of zeolite A supplementation in the dry period of periparturient calcium, phosphorus and magnesium homeostasis. Journal of Dairy Science. 2002;**85**(7):1855-1862

[41] Katsoulos PD, Panousis N, Roubies N, Christaki E, Arsenos G, Karatzias H. Effects of long-term feeding of a diet supplemented with

clinoptilolite to dairy cows on the incidence of ketosis, milk yield and liver function. The Veterinary Record. 2006;**159**(13):415-418

[42] Uyarlar C, Akkaya AB, Gultepe EE. Effects of yeast (*Saccharomyces cerevisiae*) and clinoptilolite administration on milk yield and some metabolic parameters in early lactation dairy cows. Journal of Turkish Veterinary Medical Society. 2018;**89**(1):3-10

[43] Alshannaq A, Yu JH. Occurrence, toxicity, and analysis of major mycotoxins in food. International Journal of Environmental Research and Public Health. 2017;**14**(6):632. DOI: 10.3390/ijerph14060632

[44] Aleksandra D, Magdalena T-C, Dondur V, Vujakovic A, Radosevic P. Kinetics of aflatoxin B1 and G2 adsorption on caclinoptilolite. Journal of the Serbian Chemical Society. 2000;**65**(10):715-723

[45] Safameher A. Effects of clinoptilolite on performance, biochemical parameters and hepatic lesions in broiler chickens during aflatoxosis. Journal of Animal and Veterinary Advances. 2008;7:4

[46] Miazzo R, Rosa CAR, De Queiroz Carvalho EC, Magnoli C, Chicchiera SM, Palacio G, et al. Efficacy of synthetic zeolite to reduce the toxicity of aflatoxin in broiler chicks. Poultry Science. 2000;**79**(1):1-6

[47] Goodarzi M, Modiri D. The use of clinoptilolite in broiler diet to decrease of aflatoxin effects. In: 2011 International Conference on Asia Agriculture and Animal; 13. 2011. pp. 38-43

[48] Nesic VD, Marinkovic DM, Nesic KD, Resanovic RD. Examination of the efficacy of various feed additives on the pathomorphologocal changes in broilers treated with T-2 toxin. Zbornnk Matice srpske za prirodne nauke. 2009;**116**:49-54. DOI: 10.2298/ZMSPN0916049N

[49] Rizzi L, Simioli M, Roncada P, Zaghini A. Aflatoxin B1 and clinoptilolite in feed for laying hens: Effects on egg quality, mycotoxin residues in livers, and hepatic mixed-function oxygenase activitites. Journal of Food Protection. 2003;**66**(5):860-865

[50] Katsoulos PD, Karatzi MA, Boscos C, Wolf P, Karatzias H. Field evaluation of clinoptilolite feeding efficacy on the reduction of milk aflatoxin M1 concentration in dairy cattle. Journal of Animal Science. 2016;**58**:24. DOI: 10.1186/s40781-016-0106-4

[51] Canovic MT, Dakovic A, Matijasevic S, Mihajlavic AR, Adamovic M, Stojsic D. Minazel plus—The efficient adsorbent for mycotoxins. Zbornik naucnih radova. 2003;**9**:343-351

[52] Minazel-plus [Internet]. Available from: www.patent-co.com [Accessed: 06 April 2019]

[53] Nesic VD, Ostojin MV, Nesic K, Resanovic RD. Evaluation of the efficacy of different feed additives to adsorbe T-2 toxin in vitro. Zbornnk Matice srpske za prirodne nauke. 2009;**116**:55-59. DOI: 10.2298/ZMSPN0916055N

[54] Pavelic K, Hadzija M, Bedrica L, Pavelic J, Dikic I, Katic M, et al. Natural zeolite clinoptilolite: New adjuvant in anticancer therapy. Journal of Molecular Medicine. 2001;**78**(12):708-720

[55] Pavelic K, Katic M, Sverko V, Marotti T, Bosnja B, Balog T, et al. Immunostimulatory effect of natural clinoptilolite as a possible mechanism ot its antimetastatic ability. Journal of Cancer Research and Clinical Oncology. 2002;**128**:37-44

[56] Zarkovic N, Zarcovik K, Kralj M, Borovic S, Sabolovic S, Blazi MP,

et al. Anticancer and antioxidative effects of micronized zeolite clinoptilolite. Anticancer Research. 2003;**23**(2B):1589-1595

[57] Katic M, Bosnjak B, Gall-Troselj K, Dikic I, Pavelic K. A clinoptiolite effect on cell media and the consequent effects on tumor cells in vitro. Frontiers in Bioscience. 2006;**1**(11):1722-1732

[58] Ivkovic S, Deutsch U, Silberbach A, Walraph E, Mannel M. Dietary supplementation with the tribomechanically activated zeolite clinoptilolite in immunodeficiency: Effects on the immune system. Advances in Therapy. 2004;**21**(2):135-147

[59] Sverko V, Sandra S, Balog T, Colic M, Tatjana M. Natural micronised clinoptilolite mixtures with *Urtica dioica* L extract as possible antioxidants. Food Technology and Biotechnology. 2004;**42**(3):189-192

[60] Valpotic H, Terzic S, Vince S, Samardzija M, Turk R, Lackovic G, et al. In feed supplementation of clinoptilolite favourably modulates intestinal and systemic immunity and some production parameters in weaned pigs. Veterinární Medicína. 2016;**61**(6):317-327. DOI: 10.17221/175/2015-VETMED

[61] Karatzia M. Effect of dietary inclusion of clinoptilolite on antibody production by dairy cows vaccinated against *Escherichia coli*. Livestock Science. 2010;**128**(1):149-153

[62] Fratrić N, Stojić V, Rajcić S, Radojičić B. The effect of mineral adsorbent in calf diet colostrum on the levels of serum immunoglobulin G, protein and glucose. Acta Veterinaria. 2007;**57**:169-180

[63] Marc Zarcula S, Mircu C, Bonca G, Kirovski D, Otava G, Ahmai-Khoie M, et al. Effects of clinoptilolite addition to colostrum on the concentration of serum proteins, minerals, enzyme activities in neonatal calves. Journal of Advances in Chemistry. 2014;**10**(2):2290-2296

[64] Zarcula S, Tulcan C, Cernescu H, Mircu C, Anca LN, Urian-Suesca RI, et al. Influence of clinoptilolite on some mineral parameters in neonatal calves. Lucrări Ştiinłifice Medicină Veterinară. 2011;**XLIV**(2):41-48

[65] Weaver DM, Tyler JW, Van Metre DC, Hostetler DE, Barrington GM. Passive transfer of colostral immunoglobulins in calves—Review. Journal of Veterinary Internal Medicine. 2000;**14**:569-577

[66] Samanc H, Danijela K, Adamovic M, Vujanac I, Natalija F, Prodanovic R. Effects of natural zeolite on body weight, weight gain, hematology and biochemical blood parameters in calves. Veterinarski Glasnik. 2008;**62**(3-4):153-166

[67] Marc S, Kirovski D, Mircu C, Hutu I, Otava G, Paul C, et al. Serum protein electrophoretic pattern in neonatal calves treated with clinoptilolite. Molecules. 2018;**23**(1278):1-9. DOI: 10.3390/molecules23061278

[68] Bgatova NP. Influence of long-term administration of naturally occuring sorbents on the ultrastructure of small intestine enterocyte in rats. Bulletin of Experimental Biology and Medicine. 1998;**6**:626-629

[69] Zarcula S. Research regarding immune status in newborn calves [thesis]. Romania: Banat's University of Agricultural Sciences and Veterinary Medicine "King Michael I of Romania" Timişoara; 2010

[70] Zarcula S, Cernescu H. Influence of clinoptilolite on health status in calf neonatal period. In: Proceedings of the XIth International Symposium "Young

People and Multidisciplinary Research"; 12-13 November 2009; Timişoara, România. 2009. pp. 195-200. ISSN: 1843-6609

[71] Ural DA, Ural K. Effects of short term clinoptilolite supplementation on weight gain in Holstein calves. Revista MVZ Córdoba. 2016;**22**(1):5631-5637. DOI: 10.21897/rmvz.923

[72] Rodriguez-Fluentes G, Barrios MA, Iraizoz A, Perdomo I, Cedre B. Enterex: Anti-diarrheic drug based on purified natural clinoptilolite. Zeolites. 1997;**19**(5-6):441-448

[73] Lamprecht M, Bogner S, Steinbauer K, Schuetz B, Breiberger JF, Leber B, et al. Effects of zeolite supplementation on parameters of intestinal barrier integrity, inflammation, redox biology and performance in aerobically trained subjects. Journal of the International Society of Sports Nutrition. 2015;**12**(40). DOI: 10.1186/s12970-015-0101-z

[74] Wu Y, Wu Q, Zhou Y, Ahmad H, Wang T. Effects of clinoptilolite on growth performance and antioxidant status in broilers. Biological Trace Element Research. 2013;**155**:228-235. DOI: 10.1007/s12011-013-9777-6

[75] Yildirim O, Turker A, Senel B. Effects of natural zeolite (clinoptilolite) levels in fish diet on water quality, growth performance and nutrient utilization of tilapia (*Tilapia zillii*) fry. Fresenius Environmental Bulletin. 2009;**18**(9):1567-1571

[76] Hcini E, Slima AB, Kallel I, Zormati S, Traore A, Gdoura R. Does supplemental zeolite (clinoptilolite) affect growth performance, meat texture, oxidative stress and production of polyunsaturated fatty acid of Turkey poults? Lipids in Health and Disease. 2018;**17**:177. DOI: 10.1186/s12944-018-0820-7

[77] Martin-Kleiner I, Flegar-Meštric Z, Zadro R, Breljak D, Stanović Janda S, Stojković R, et al. The effect of the zeolite clinoptilolite on serum chemistry and hematopoiesis in mice. Food and Chemical Toxicology. 2001;**39**:717-727

[78] Guilloteau P, Zabielski R, Blum JW. Gastrointestinal tract and digestion in the young ruminat: Ontogenesis, adaptations, consequences and manipulations. Journal of Physiology and Pharmacology. 2009;**60**:37-46

[79] Pârvu G. Animal Metabolic Surveillance. Bucharest, Romania: Ceres; 1992

[80] Marc Zarcula S, Tulcan C, Kirovski D, Samanc H, Mircu C, Bonca G, et al. Effects of short term addition of clinoptilolite to colostrum on some biochemical parameters in newborn calves. Lucrări Ştiinłifice Medicină Veterinară. 2013;**XLVI**(3):167-172

[81] Zarcula S, Tulcan C, Mircu C, Kirovski D, Samanc H, Cernescu C. Influence of clinoptilolite on some enyzmatic parameters in neonatal calves. In: Proceedings of the XIIth International Symposium "Young people and multidisciplinary research"; 11-12 November 2010; Timişoara, România. 2010. pp. 167-173. ISSN 1843-6609

[82] Boyd JW. Serum enzyme changes in newborn calves fed colostrum. Veterinary Clinical Pathology. 1989;**18**(2):47-51

[83] Andrei S, Groza IS, Pintea A, Ciupe S, Chirila F, Miclean M. Physiology and Pathology of the Mammary Gland in the Cow. Cluj: Academic Press; 2010. ISBN: 978-973-744-232-1